Anthony M. Valeri, MD

NYU-Polytechnic University, BS (major: Life Sciences), 1973, Summa Cum Laude

SUNY- Health Science Center at Brooklyn, MD, 1981, Magna Cum Laude

Residency in Internal Medicine, New York Presbyterian Hospital (Columbia), 1981-1984

Clinical Fellowship in Nephrology, New York Presbyterian Hospital (Columbia), 1984-1986

Assistant Professor of Clinical Medicine, NY University -Bellevue, 1986-1991

Professor of Clinical Medicine, Columbia University College of Physician and Surgeons

New York Presbyterian Hospital (Columbia), 1991-Present

Medical Director, Hemodialysis, New York Presbyterian Hospital (Columbia), 1991-Present

Physician of the Year Awardee, New York Presbyterian Hospital (Columbia), 2001

Hobby: Reading and sightseeing.

Kyung-Sook Kim, RN, MSN, CNN, FNP- C

New York Presbyterian Hospital

Gyeong Gi Nursing School, Incheon, Korea

Bloomfield College, BSN, Cum Laude

Long Island University, Master of Science in Nursing

Certified Nephrology Nurse

Board Certified Family Nurse Practitioner

Member of American Association of Nurse Practitioner

Hobby and specialty: playing piano & harp, reading and writing (writing "Thank You" letter)

Acknowledgements

Robert Kelly, MD
President of New York Presbyterian Hospital

Anthony Valeri, MD
Professor of Clinical Medicine
Columbia University College of Physicians and Surgeons

Jai Radhakrishnan, MD
Professor of Clinical Medicine
Columbia University College of Physicians and Surgeons

Robin Ferrer. RN. MSN
Patient Care Director of Hemodialysis unit
New York Presbyterian Hospital

Thank for your support and help.
In addition to this, thank you for teaching me the beauty of modesty.

Dr. Valeri, Thanks for teaching me very nicely for several years with your enormous knowledge in medicine. I deeply appreciated it.

My beloved family, Jae-kwon, Enoch and Eunice. Thank you for love, help and patience. I deeply appreciated it.

Dear God,
Thank you. I love you with all my heart.
Soli Deo Gloria! (Only for the glory of God)

Kyung-sook Anne Kim

Special Thanks to Daphney Powell

My head nurse, mentor, counselor and mother. God sent you to me because I miss my mother so much. It is a true blessing from heaven above that I have you in my life.

I deeply appreciated your care, love, concern and encouragement for the past 25 years.

I love you Daphney.

Kimmie (Kyung-Sook Kim)

"A productive healthcare-provider relationship is built on the foundation of excellent communication. I had the privilege to review Ms. Kim's book, "English for Healthcare professionals", and this book is an important contribution in this area. Ms. Kim's draws upon her extensive clinical experience in an academic New York City teaching hospital which takes care of patients from many cultures and who speak many different languages. She has drawn up a list of common clinical encounters where the healthcare provider is taking a history and then providing counseling to a patient. The list of diseases is quite comprehensive and each topic is dealt with in-depth. They body of the discussion is true to life and brings out the medical subtleties of the English language during these encounters. This Book is an excellent resource for Medical English for practitioners whose first language is not English and will go a long way in building a sound patient-provider relationship."

Jai Radhakrishna, MD, MS, MRCP, FACC, FASN
Professor of Clinical Medicine
Columbia University College of Physicians and Surgeons
Associate Clinical Director and Director of the Nephrology Fellowship Program
New York Presbyterian Hospital

"I enjoyed reading this fabulous book by Kim. It brought me right to the wards and clinics in US hospitals. With real cases, real settings and real dialogues, the book is very practical and interesting to read. It teaches me just what is needed in US hospitals. As an ESL healthcare professional from China, this book does help me tremendously in adjusting to the US healthcare system, not only language wise but also knowledge and culture wise. Also, I believe it is a perfect supplemental study material for the USMLE CS exam. I am sure you will enjoy reading this book as I do!"

Weijia Wang, MD, China

"I fell in love with the book as soon as I read Ms. Kim's first paragraph… As a non-English native speaker, I found *English for Healthcare Professionals* to be wonderful, helpful and pleasant to read. This friendly book truly reflects what the most important fact is in a clinical encounter: communication with the patient; and it guides the reader with great detail, through the clinical scenario emphasizing the medical vocabulary, and the nature of the dialogue in a given situation.

It is undoubtedly a clever, precise, friendly and complete tool that every non-English native speaker in the process to become a healthcare professional in the U.S. should read. Thanks dear Kim for putting this wonderful book on my way…"

Angelica Cifuentes, MD, Colombia

Introduction of "English for Healthcare Professionals"

When I was in graduate school training to become a family nurse practitioner, I wrote many practice notes that followed the "SOAP (subjective, objective, assessment and plan)" format. It was a lot of hard work, but eventually I realized how valuable they were when I was actually interviewing patients.

One day last summer, I had a chance to review my notes and I thought this would be good resource for those healthcare providers for whom English is a second language (ESL), to help them improve their medical English.

This book is written based on true clinical scenarios and consist of actual conversation which will occur between patients and doctors in a U.S. hospital.

Readers can learn not only conducting a conversation in medical English but also understand diagnostic strategies and other clinical aspects of each disease as is done in American hospitals.

I am sure that ESL healthcare professionals who are interested in learning American medical English will benefit by reading this book. I learned my medical and communication skills from one of the best hospitals in the world (New York Presbyterian Hospital), and would like to share this knowledge with other professionals in the healthcare field.

Contents

1. Allergic Rhinitis

CC: "I have a runny nose and my eyes are itching".

Rona is a 55 y/o Caucasian female who presented with a runny nose, itching, tearing eyes and sneezing for 3 days. Her symptoms got worse after she went out for her doctor's appointment. She immigrated to the United States from Spain when she was 25 years old with her family and she developed allergies 7 years later for the first time. Her allergy symptoms included severe itching in her eyes and constant sneezing at that time.

The itching in her eyes was so severe that she thought she would scratch her eyeballs out. She went to her PCD and took some medications he prescribed, but she felt only temporarily relief. She felt much better with a change of weather.

Her allergic symptoms have recurred every spring thereafter. Her doctor performed allergy skin tests and found that she is allergic to pollen. She wanted to stay home because of her history, but she had to go to see her doctor for her chronic back pain.

In addition to these symptoms, she also c/o headache, sore throat, mild periorbital puffiness along with a clear nasal discharge.

She denied fever, chills or SOB (shortness of breath). She was diagnosed with hypertension 6 years ago and she is on HCTZ (hydrochlorothiazide) and Quinapril.

She lives in a private house with her husband. She was a teacher's aide but she retired early to help take care of her grandchildren.

Dialogue

Doctor: Hi, Rona. What brings you in today?

Rona: I have a runny nose and my eyes are itching very badly.

Doctor: When did you first notice these symptoms?

Rona: I have had them for about 3 days, but it is like I am having it every spring.

Doctor: Do you have any other symptoms?

Rona: I sneeze too, but the itching is really driving me nuts. I am afraid I might scratch out my eyeballs.

Doctor: You said it happens every spring?

Rona: Yes, according to my doctor, I am allergic to pollen so my allergy symptoms get worse during springtime always. This time, my symptom started after I went out to see my doctor because of my back pain.

Doctor: Do you have nasal stuffiness?

Rona: No, but I think I have a little swelling around my eyes.

Doctor: Have you ever received any treatment for your symptoms in the past?

Rona: Yes, actually, my allergies developed about 7 years after I came to this country. I saw a doctor at that time and he prescribed some medication, but I felt better for only a few days with the medication. When the weather gets warmer, I feel better.

Doctor: What is the color and amount of your nasal discharge?

Rona: It is clear, but it is running constantly.

Doctor: Is there any changes in your hearing or sense of smell?

Rona: No, I think my hearing and smell are fine.

Doctor: Do you have a sensation of needing to constantly clear your throat?

Rona: No, I am O.K. My major problems are runny nose and severe itching in my eyes.

Doctor: Do you have headache or a sore throat?

Rona: Yes, both sometimes.

Doctor: Any of your family members with the same problem?

Rona: One of my sisters has the same allergy problem like me after we moved to the States.

Doctor: Let me examine you. I think your allergic rhinitis has recurred this spring. In your case, avoidance of the allergen is the most effective form of treatment. I will prescribe a steroid nasal spray to help reduce the inflammation that will help alleviate your symptoms. If that doesn't help, we could try a series desensitization shots. I would refer you to an allergy specialist for that.

2. Angina Pectoris

CC: "I get tightness in my chest."

Sarah is a 65 year old Caucasian female who presented with chest pain and tightness while gardening earlier today. She described her pain as pressure, squeezing, burning and tightness in the chest. Her pain started behind the breast bone and radiated up to her neck. This pain occurred when she was working in her garden in the middle of the day after she had a heavy lunch.

The pain lasted for 5 - 10 minutes then resolved when she rested inside her home. Her chest pain was graded 6 - 7 out of 10. This pain was accompanied by SOB (shortness of breath) but no sweating, nausea or vomiting. She did not attempt any other measure to relieve her pain but rest.

She describes no other associated symptoms during this episode of pain such as dizziness or palpitations. Sarah initially thought she had indigestion, but she has decided to seek medical attention when a second episode of pain occurred 3 hours ago. At the clinic, she had another episode of chest pain and tightness and was given NTG (nitroglycerin) 0.4 mg sublingually.

Her pain resolved within 5 min after the administration of SL (sublingual) NTG. She had been diagnosed with DM (diabetes mellitus) 15 years ago and HTN (hypertension) 10 years ago.

She stated that her cholesterol is high but she does not know how high it is. She does not have any GI (gastrointestinal) problem history in the past. Sarah's father passed away at the age of 52 due to a heart attack and her mother has had diabetes for over 20 years.

Dialogue

Doctor: Sarah, How do you feel today?

Sarah: I feel horrible. I have tightness in my chest and I am little bit scared of my symptoms.

Doctor: When and how did it happen?

Sarah: I was working in my garden all morning and afternoon. I felt chest pain and tightness after I had large meal because I was starving. It happened while I was gardening to plant some new flowers.

Doctor: How long does it last?

Sarah: I think it lasted roughly for about 5 -10 minutes.

Doctor: What did you do to relieve your chest pain? Did you rest?

Sarah: Yes, I went into my house and sat in a chair for a while until the pain went away.

Doctor: How strong was your pain? If you use a scale of 1-10, 1 as the mildest and 10 as the strongest pain you can imagine. How would you grade your pain?

Sarah: I guess my pain was 6 or 7 out of 10. It was hurting a lot.

Doctor: Did you have any other symptoms such as shortness of breath, sweating, nausea, or vomiting along with the chest pain?

Sarah: I had SOB, but I did not have any other of the symptoms you mentioned. Actually, I initially thought I had indigestion because I just had a big meal, but the chest pain came back 3 hours ago and I have decided to check with you.

Doctor: Does anyone in your family have heart problems, diabetes or hypertension?

Sarah: Yes, my father passed away from a heart attack and my mom has diabetes.

Doctor: Sarah, You have several risk factors for developing heart disease, your diabetes, high blood pressure, age, and your family history. Let's investigate your symptoms further to prevent your chest pain episodes in the future. I want to get some blood work (CBC, BUN/ Creatinine, Troponin, CPK, Hb A1c, ESR/CRP, Lipid panel) and an EKG today. I will give you a prescription for Aspirin and for *NTG to take if you get another episode of the same pain. You can take up to 3 pills, 1 every 5 minutes, if the pain doesn't go away; call 911 immediately to take you to the nearest ER (Emergency Room). Otherwise we'll plan for a stress test and echocardiogram to evaluate for the presence and severity of coronary artery disease and to evaluate left ventricular function. Based on those results, we would begin aspirin, beta-blockers and statin therapy. Cardiac catheterization may be indicated based on the severity and extent of the coronary artery disease suggested by the stress test and echocardiogram.

*NTG: Nitroglycerin

3. Arrhythmia

CC: "I feel dizzy, weak and I have palpitations."

Wong is a 68 Asian male who came to the clinic for dizziness, palpitations and weakness on occasion and when he had an episode most recently, he nearly fainted. After this most recent episode, he became so scared about his condition that he wanted to see his doctor sooner than his routine scheduled appointment time. He was diagnosed with diabetes when he was 63 years old and he is on metformin 1000mg twice a day and his blood sugar has been well controlled. He is a smoker and he smokes about a half pack per day for the past 30 years and he consumes alcohol heavily with his friends most weekends. When his wife complained about his smoking and drinking habits, he was very upset and did not come home for 3 days. He stayed at his brother's home and Wong's wife became very concerned about her husband. He told his wife "I am old enough to do what I want to do. Just leave me alone." Wong thought his life was so consumed taking care of his family that he now deserved to do what he wanted to do for himself.

He reduced his work hours by almost half because he is thinking about retiring soon to enjoy his life with his friend and doing things he could not do before because of his multiple responsibilities and domestic obligations as the bread winner of his family. He wanted to retire at the age of 65 but he was not able to carry out his wish due to the depressed economy and he was very stressed out about it. Both of his parents have hypertension for many years and one of his brothers has hypertension and diabetes like Wong. Wong loves to drink coffee and he consumes 3 - 4 cups of coffee every day.

Dialogue

Doctor: How are you, Wong?

Wong: I do not feel well. I feel dizzy, weak and I have palpitations at times.

Doctor: How often do you feel these symptoms?

Wong: I do not remember clearly but I get them at least 3-4 times a month. When I felt these symptoms the last time, I thought I was going to pass out. It was a horrible feeling. Actually, I got pretty scared about it.

Doctor: Do you have any chronic medical problems such as hypertension or diabetes?

Wong: Yes, I have both.

Doctor: Did you have any chest pain along with your symptoms?

Wong: No.

Doctor: Your symptoms, how long do they last?

Wong: Because they make me feel so scared, it seems like a very long time but I think they really last just a couple of minutes.

Doctor: Do you drink alcohol or smoke cigarette?

Wong: Yes, my wife keeps nagging me a lot about this so I went to stay at my brother's house for a couple of days to get away from her nagging.

Doctor: Wong, that's not nice. She must be so worried about you.

Wong: Doc, I am old enough to do what I want to do. I will not allow anybody to harass me anymore.

Doctor: Do you drink coffee? If so, how many cups of coffee do you drink per day?

Wong: I love coffee. When I open my eyes in the morning, I make my coffee. Coffee is an eye opener for me. After I have a cup of coffee, I start my day. I think I consume 3-4 cups of coffee every day.

Doctor: Do you notice any mood swings or any changes in your weight or bowel habits?

Wong: Not really. I do not think so.

Doctor: Do you have any heat or cold intolerance or shakes in your hands?

Wong: No.

Doctor: Tell me about your life. Do you have any stress in your life? I know everyone has stress in their lives including myself.

Wong: (with deep sigh) Actually, I wanted to retire 3 years ago at the age of 65 but I was not able to do it because the economy is so depressed. I worked all my life as an engineer for my family. I am a first son of my family. Doc, do you know the meaning of the first son of the family in my culture?

Doctor: Sorry, I do not know but I can assume more responsibilities for your family, right?

Wong: Exactly, the first son has to take care of their parents. It is not a law but our culture expects that taking care of elderly parents is the first son's duty and responsibility. I take care of my parents as much as I can. I love my parents and I wanted to please them so I tried my best but my good intentions were a heavy burden on my shoulder on many occasions honestly. I need a break for myself and I want to do something that pleases me instead of working every day. I am trying to retire this year but I do not know when I can do it. I am already 68 and I am tired.

Doctor: OK. I see. Is there anybody in your family with heart disease?

Wong: No, but both of my parents have hypertension. What is going on with me? What is the cause of my symptoms, doc?

Doctor: I think you are having attacks of your heart beating rapidly and I have to investigate your symptoms further. A rapid heart rate can come on suddenly and usually from the atria, the top chamber of the heart, (SVT - supraventriculatr tachycardia) and can be regular (PAT or PSVT sometimes triggered by an electric pulse going through a "short circuit "path in the heart or atrial flutter)or irregular (PAF-paroxysmal atrial fibrillation). It can be due to an intrinsic heart problem or some other stimulus outside the heart such as from excessive caffeine or alcohol intake (holiday heart syndrome) or an overactive thyroid (hyperthyroidism). Intrinsic heart disease like coronary artery atherosclerosis or valvular heart disease (aortic or mitral valve stenosis or insufficiency) or disease of the heart muscle (intrinsic or alcoholic cardiomyopathy) can also cause the problem.

Wong: Do I have to take special tests?

Doctor: You will need several tests to confirm the diagnosis. The testing we should do include:

1. An electrocardiogram

2. An echocardiogram - to check for valvular problems and assess the size of each chamber of the heart.

3. A holter monitor to try to identify the abnormal heart rhythm when it occurs.

4. Thyroid function tests

5. Exercise stress test to evaluate for coronary artery disease.

Based on the test results, additional tests such as cardiac catheterization (to assess the coronary blood vessels for blockage and the heart valves for stenosis or insufficiency) or an EPS (electrophysiology studies) - to assess for abnormal pathways of electrical conduction in the heart maybe necessary. Abstinence from smoking and alcohol to the extent possible is vital along with cutting back on caffeinated beverages. If the thyroid is overactive, we would refer you to an endocrinologist for further testing and treatment.

Abnormal heart rhythms can often be controlled with medication such as beta-blockers, AV nodal blocking calcium channel blockers, or amiodarone.

But some patients require interventions such as ablation of an abnormal conduction pathway if found on EPS.

4. Arthritis

CC: "I have pain and stiffness in my joints and it has been getting worse for the last 2 months".

Judy is a 75 year old Asian female who presented to the clinic with worsening joint pains and stiffness. Judy does not recall when this pain started for the first time, but she has had stiffness and joint pain in both her hands and right knee. The intensity of her pain has been getting worse, especially in her knee when she gets up after sitting down for prolonged periods of time. Judy loves to knit and has spent a lot of time knitting sweaters in order to give them as gifts to her close friends during the holiday season. She noticed one day while knitting that she felt pain in her hands that was much worse compared to other days, and therefore, has tried to reduce the time she spends knitting.

Judy's stiffness gets better with using her hands and lasts about 20-30 minutes early in the morning. She took Acetaminophen (Tylenol®) to relieve her pain for some days, but it did not help as much as she expected. Judy also has used a heating pad to relieve the pain in her knee. She is a moderately overweight, well-nourished female who has mild discomfort when arising from sitting down. Her joint pain and stiffness has gradually increased over the past 3 years and she recently realized that she has developed ugly bumps on her finger joints, even though they are not painful. Nonetheless, Judy is concerned about her finger joints. She does not have any allergies to foods or drugs. She lives on the first floor of a private house with her husband and one adult son.

The intensity of her pain affects her daily activities significantly.

Dialogue

Doctor: Judy, what brings you in today?

Judy: My joints have been stiff and painful for about 3 years. It was tolerable with my pain killers, but now the pain is getting progressively worse.

Doctor: About how long does your joint stiffness last?

Judy: It lasts about 20 - 30 minutes in the morning after I get up from bed, and I feel better with using my hands.

Doctor: Do you have any warmth or swelling along with your stiffness in your joints?

Judy: No.

Doctor: When do you feel more pain? Are there any aggravating factors that you can recall?

Judy: Yes, one of my hobbies is knitting and when I spend my time knitting, I feel much more severe pain. But I just love to do it.

Doctor: How about any other activities such as prolonged walking, standing, or gardening? I know you love gardening too.

Judy: Yes, that is right. I feel discomfort and pain when I get up after sitting for a long time, and after kneeling a long time while gardening.

Doctor: Do you take any medications to help relieve your pain?

Judy: Yes, I take Acetaminophen (Tylenol®) occasionally.

Doctor: How often and how many tablets do you take?

Judy: I try not to take medication if possible. I take them only when I feel intense pain. When I take medications, I take 2 tablets at once.

Doctor: Does your pain affect your day-to-day activities significantly?

Judy: Yes, because of the joint pain, I can't even walk 10 blocks without much difficulty, and the stiffness of my joints prevents me from doing certain finger movements. Do you also see the bumps on my hands? I do not like it. Do I have to see a surgeon to get rid of these bumps?

Doctor: No, surgical intervention for the removal of these bumps is not necessary because they are not harmful. We'll get some x-rays of your hands and series of a special blood tests, a rheumatoid factor/latex fixation test, a CCP antibody, (anti-cyclic citrullinated peptide antibody) and ANA (anti-nuclear antibodies) test to determine what kind of arthritis you have. It might be rheumatoid arthritis, but will also test for other connective tissue-collagen vascular diseases like SLE (systemic lupus erythematous) and MCTD (mixed connective tissue disease). We'll try a stronger, longer lasting pain and anti-inflammatory medication. It is called Naprosyn and it's in a family of medications called NSAID's (Non-Steroidal Anti Inflammatory Drugs). Also, I will send you to a rheumatologist.

5. Asthma

CC: "Tom could not sleep last night because of wheezing and shortness of breath."

Tom is a 7 year old boy who has a long history of asthma. As an infant, Tom suffered from otitis media several times which resolved with antibiotics. At the age of 3, Tom had an episode of persistent, nonproductive coughing which worsened at night. He was then diagnosed with asthma soon thereafter.

Tom visits the emergency room every three to four months due to severe wheezing and chest tightness along with shortness of breath that worsens at night, and his symptoms were not controlled with his medications.

Tom's doctor performed a skin test on him and according to the results of his test, Tom is allergic to carpet dust, cat, and dog dander. He was also exposed to second hand smoke at home because his father is a smoker, but his father tries not to smoke around him. However, Tom's father has admitted to smoking in the house before Tom was diagnosed with asthma. Tom's father has tried very hard to quit smoking, and he had counseling for his smoking cessation. But it has been really difficult for him to quit smoking because he has been smoking since he was 17.

Tom lives in a wall-to-wall carpeted house with his parents and his two younger siblings.

His mom also has a cat and dog. Tom's mom informed the doctor that Tom loves these animals and doesn't want to give them up.

They were told that the animals should be kept outside as much as possible, and Tom should not touch these animals even if he loves them. Tom was instructed to stay away from these animals after the results of the skin test, but Tom and his parents were not compliant with this recommendation.

Dialogue

Doctor: Peter, how is Tom?

Peter: My son is coughing a lot, especially at night. Tom could not sleep last night because of the coughing and couldn't breathe well. I was so scared.

Doctor: Did he have this problem before? Does he bring up anything when he coughs?

Peter: Yes, we had to bring him to the emergency room several times in the past because of the same breathing problem. He also complained about chest tightness and I could hear him wheezing. He did not cough up any phlegm though, and just coughs persistently.

Doctor: Has anyone told you Tom has asthma?

Peter: Yes, Tom was actually diagnosed with asthma when he was 3.

Doctor: Does he take any medications to control his asthma?

Peter: Yes, but his asthma has not been controlled well with his asthma medications and I am so worried whenever he coughs a lot at night.

Doctor: Was he ever exposed to second-hand smoking?

Peter: (pause)……. Yes, I smoke cigarettes but I have tried very hard to quit smoking. I am thinking about joining a smoking cessation class.

Doctor: That's a good idea. Has anybody performed skin testing on Tom?

Peter: Yes, his doctor did skin tests on him to see if he is sensitive to anything. I was informed that Tom has a sensitivity to carpet dust, dog dander, and cat. Doctor, to be honest with you, we have a cat and dog at home. All of us are animal lovers and Tom is especially crazy about pets and we do not want to give them up even though we know they aren't good for Tom.

Doctor: If you can't get rid of your pets, you'll have to raise them outside of the house. Those pets should also stay outside of Tom's room. For asthma, environmental factors are critical. Even if Tom loves these pets, he is not allowed to touch them at home. Can you keep a journal to monitor his symptoms and medication use? I will review this journal at his next visit to check the overall efficacy and response to treatment. Also, I will prescribe a long acting bronchodilator plus a rescue inhaler for acute worsening of his symptoms. I also recommend a series of desensitization injections if you are unwilling or unable to remove the environmental triggers such as your pets.

6. Back pain

CC: "I have had low back pain for about 3 weeks and it does not get better no matter how hard I tried to get rid of it."

Susan is a 45 year old Hispanic female who works in the pantry of a large rehabilitation center. Susan has had low back pain for about 3 weeks after she tried to lift a heavy vegetable box on her own due to a staff shortage. Usually, that box was moved by male employees, but her pantry was extremely busy on that day, and she was not able to get any help. Right after she moved the heavy vegetable box, she started to feel the pain and her pain has been getting worse for the last 10 days. She tried to rest and applied a hot pack on her painful back to alleviate the pain, but it did not help to relieve her pain at all. She even tried an over- the-counter painkiller, but the pain persists and has reached the point where she can't bear it anymore. She feels pain constantly in her lower back and it radiates down her right leg. She was limping when she tried to walk due to her pain and on a pain scale, it is six or seven out of ten. She had another episode of low back pain about 6 months ago and she had to stay out of work for 2 weeks until she felt better.

Susan is a single mom who raises 3 children all by herself. She is deeply concerned about her low back pain because she has already used up all her sick time for this year. She stated that if she has to stay out of her work again, she will face serious financial problems. Because of her financial issues, she forced herself to continue working, but she visited the clinic today to seek medical advice because her pain is unbearable now. Her three children are in middle school and high school, and they are doing fine in school. Both of her parents are healthy and they try to help her whenever they can. Susan has two brothers and one sister. They live in another state and they do not have any serious health issues.

Dialogue

Doctor: How are you Susan? What brings you in today?

Susan: I have had severe back pain for about 3 weeks.

Doctor: How did it start? Did you do something out of the ordinary?

Susan: I tried to move a heavy vegetable box on my own because of a staff shortage and I started to feel the pain right after I had finished moving the box. It was really heavy, but I wasn't able to get any help moving it because everyone else was busy with their own assignments. But, now I regret that I did not wait till I got some help. I should have remembered that I had the same back pain just 6 months ago.

Doctor: Please tell me about the nature of your pain. Is it pressing or pinching?

Susan: I feel like something heavy is pressing on my back constantly, and it is really painful. I also feel the pain shooting down by right leg.

Doctor: Is there anything aggravating or alleviating your pain?

Susan: I tried everything I can do at home. I rested my back for a couple of days, then I applied a hot pack over it on top of taking Acetaminophen (Tylenol ®)to relieve my pain, but none of it seemed to help my pain. I feel miserable today.

Doctor: You mentioned you had back pain just 6 months ago. Was it the same kind of pain?

Susan: Yes, almost the same except the intensity of the back pain is much more severe now than the last time it occurred.

Doctor: Did you report your injury to your boss? Your immediate boss should report your injury to your employer when you return to work. I will write a note for your employer. We will give you a stronger pain medication and a muscle relaxant and see how you are in 3 weeks. If you do not get better, we'll schedule an MRI (magnetic resonance imaging) to see if you have herniated disc or pinched nerve in your back. Come back immediately if you develop weakness in the leg or have trouble passing urine or stool. We should also send you for a DEXA (dual emission X-ray absorptiometry) to check your bone density and check your vitamin D levels (25-hydroxy vitamin D). You may benefit from calcium and vitamin D supplementation.

7. Bacterial Vaginosis

CC: "I have a vaginal discharge and it has a fishy smell."

Crystal is a 38 y/o Hispanic female who visited GYN clinic because of a vaginal discharge for about 6-7 days. The color of the discharge is gray and it has a foul smell and it smells like fish.

She stated that she wore a strong perfume to cover the fishy smell because she felt embarrassed going out in public.

Patient denies vaginal itching or burning on urination. Patient is sexually active and she is in a monogamous relationship with her husband of 8 years. Initially, she was suspicious that her husband must have given it to her because she had a history of STD (sexually transmitted disease) from her husband in the past. So she confronted her husband before she came to the clinic, but he denied that he had been unfaithful.

She decided to trust her husband this time and she was hoping that her symptoms would get better without seeing a doctor, but it did not get better. She even tried frequent douching to see if it would help to improve her symptoms, but it did not help at all.

Her LMP was 2 weeks ago and it was uneventful and she uses an oral contraceptive.

Dialogue

Doctor: Hi, Crystal. How are you?

Crystal: I have a foul smelling vaginal discharge and it is very embarrassing.

Doctor: Can you tell me the color of your vaginal discharge?

Crystal: It is a white to gray discharge, but the smell is horrible. I stayed outside of the waiting room until you called my name. I feel like my whole body smells so I wore strong perfume to cover the smell. Still I am not comfortable to sit with other patients in the waiting room.

Doctor: How about the amount? Is it a lot?

Crystal: Yes. It is copious.

Doctor: Do you have any other symptoms such as vaginal itching or burning on urination?

Crystal: No. Just a fishy smell that bothers me the most.

Doctor: Do you do frequent douching or use feminine hygiene products because of the odor?

Crystal: I do not use feminine hygiene products, but I did frequent douching since I developed the symptoms, but it did not help. Also, I even confronted my husband because he infected me with a STD in the past. He swore to me that he was not unfaithful and then he flipped out. He did not talk to me for 2 days because I confronted him. He is such a narrow minded man, but I think it is not because of him this time. Last time he infected me, he admitted readily with an apology when I asked him a second time.

Doctor: (With smile) Crystal, don't jump to any conclusions and blame your husband. I think your husband may be completely innocent this time. What kind of contraception do you use? Do you use any contraceptive or condom?

Crystal: My husband does not like to use a condom so I take the pill.

Doctor: Based on your history, you might have a bacterial infection of the vagina. Let me examine you. If you have a bacterial infection of the vagina, we call it Bacterical Vaginosis. If you have Bacterial Vaginosis, I will prescribe a medication called Metronidazole. You should not consume alcohol while you are on this medication and for a couple of days after you finish the medication. This is very important. Otherwise you might have a serious reaction to the alcohol. This medication can prevent the liver from metabolizing alcohol properly.

Crystal: Do I have to bring my husband in too?

Doctor: No, your husband doesn't need to come. Bacterial Vaginosis is not a STD.

8. Benign Prostate Hyperplasia

CC: "I can't sleep well because I have to get up to go to the bathroom several times every night".

Alan is a 68 year old Caucasian male who presented with a weak urinary stream, frequency, urgency, and nocturia for about one month. Alan has noticed that he has trouble getting a urine stream started and completely stopped. He has to get up several times at night to go to the bathroom because of the feeling that he needs to urinate.

However, if Alan goes to the bathroom, he has difficulty starting to urinate and his urine stream is very weak. Even after he voids, Alan has the sense that his bladder is not completely empty. This feeling of incomplete voiding makes Alan very nervous and anxious.

He cannot sleep well at night due to his frequent trips to the bathroom, and this lack of a decent amount of sleep makes him tired throughout the day. Alan has denied having a burning sensation on urination, and he also denies fevers, chills, N/V/D (nausea, vomiting, and diarrhea) or flank pain. No hematuria is reported.

Alan has had hypertension for 20 years and is on several medications to control his blood pressure. He is a retired court clerk and enjoys taking walks with his wife twice a day, but he hasn't been able to do this because he feels extremely tired. He lives with his wife and adult daughters. His wife is healthy except borderline hypertension but is not on medication.

Dialogue

Doctor: Hi Alan, you look tired. What brings you in today?

Alan: I can't sleep well at night because I need to get up and go to the bathroom so often, every single night.

Doctor: How often do you have to go to the bathroom at night?

Alan: Five or six times a night, and even if I urinate, I still have the sense of urine still in my bladder. It makes me feel nervous and anxious.

Doctor: Do you have any problem when you start to urinate?

Alan: Yes, I have trouble getting my urine started and I continue to dribble a little bit at the end.

Doctor: Have you ever noticed whether there is a change in your urine stream?

Alan: Yes, my urine stream seems weaker than before.

Doctor: Over the past month, how often do you start and stop when you urinate?

Alan: I don't remember exactly, but I call tell you it happens quite often.

Doctor: Over the past month, how often have you found it difficult to hold in your urine?

Alan: Roughly 3-4 times a day. When I feel like I have to go, I can't wait and I have to rush.

Doctor: Do you have any pain when you urinate, like any burning sensation?

Alan: No.

Doctor: Have you noticed any blood in your urine, or any flank pain?

Alan: No, the color of my urine is light yellow and I do not have any flank pain.

Doctor: Okay, Let me check your prostate and we'll get some blood tests to check your prostate, (PSA - prostate specific antigen- which can be mildly elevated in BPH or very elevated in prostate cancer), blood sugar, (for new onset of diabetes with polyuria causing nocturia) and kidney function (as CKD will impair urinary concentrating ability and cause nocturia.) We'll start a medication called an alpha blocker, which will help ease urinary flow by relaxing the urethra vascular musculature to see if that helps with your symptoms.

9. Breast Lump

CC: "I found a lump in my breast."

Margaret is a 48 year old white female who visited the clinic because she found a lump in her right breast. She found it accidently while she was showering about 1 month ago and she became very anxious because her mom is a breast cancer survivor. She could not come earlier to the clinic due to fear of getting bad news. Her lump is non-tender to the touch, soft and mobile. Her menarche was at age 16 and her LMP was 10 days ago. There is no discharge from her nipple. Margaret examines her breast monthly a week after the onset of her menses, and did not feel any lump until 1 month ago. Margaret's last mammogram was 5 years ago and she was not able to go back for another one due to her hectic work schedule and domestic duties. Margaret is married and still sexually active. She maintains a monogamous relationship with her husband of 18 years.

She uses an IUD (intrauterine device) for contraception and does not have any problems with it. She has no chills, fever, abdominal pain, or pelvic pain. Margaret used to smoke cigarettes for 25 years, but now has stopped smoking, and is a social drinker who only drinks alcohol on special occasions. Margaret's 78 year old mother was diagnosed with breast cancer when she was 45 and had a right mastectomy. Her mother is healthy and she does not have any medical problems except hypertension.

Dialogue

Doctor: How are you Margaret?

Margaret: I found a lump in my right breast accidentally about 1 month ago. I did not come earlier because I was little bit scared of my family history of breast cancer.

Doctor: I understand that. Tell me about your lump. Is it painful?

Margaret: No.

Doctor: Is it hard or soft? Is it movable?

Margaret: It is soft and I can move it around.

Doctor: Is there any discharge from your nipple?

Margaret: No.

Doctor: Did you have injury or trauma to your breast recently?

Margaret: No, not that I recall.

Doctor: Do you use oral contraceptives?

Margaret: No, I am using IUD. That is more convenient for me.

Doctor: Any problem with it?

Margaret: No, I am ok.

Doctor: Please tell me about menstrual history. Are your periods regular? When did you have your first period? What about the amount?

Margaret: My first period was when I was 16. It was kind of late and my mom worried about it so much, but my periods are pretty regular right now. I have my period for about 3-4 days and I think I bleed an average amount.

Doctor: Do you perform BSE (breast self-examination)? If you do, when do you do it?

Margaret: I do it every month about one week after my period starts as I was instructed.

Doctor: Yes, That's right. You have to do it when your breast tissue is the least congested. When was your LMP?

Margaret: My last period ended 10 days ago.

Doctor: You have a family history of breast cancer, right?

Margaret: Yes, My mom had breast cancer when she was about my age and I scared of that fact. I heard breast cancer can run in the family.

Doctor: A family history of breast cancer is certainly one of the risk factors, but that doesn't mean you are going to get it for sure. Let's investigate your lump further, but most breast lumps are benign. You had your last mammogram 5 years ago and you need your mammogram soon. I will order it today. Depending on the result, you might need further testing, an ultrasound or MRI, and it may come to needing a biopsy to find out for sure what it is.

10. Candida Vaginitis

CC: "I have a white discharge from my vagina".

Emma is 45 y/o Caucasian female who went to the gynecology clinic because of a white vaginal discharge for about 6-7 days. The discharge has no odor and is thick in consistency.

She also c/o vaginal itching and soreness along with burning on urination.

Recently, she had an upper respiratory infection and she was prescribed antibiotics for 5 days.

She is sexually active and she has been in a monogamous relationship with her husband of 15 years. She reports that she and her husband have no other sex partners.

She was diagnosed with hypertension 2 years ago and she takes a walk in the park nearby from her house 3-4 times a week for exercise to help control her blood pressure.

Emma does not have any STD (sexually transmitted disease) history and her LMP was one week ago. She does not have dyspareunia (painful intercourse).

Emma lives with her husband and two children in a private home and she likes to do gardening during her leisure time. She works in the post office and she enjoys spending time with her friends and relatives on the weekends. She likes cooking and she takes cooking lessons once a week. Emma is a non-smoker and she drinks for special occasions only with family and friends.

Dialogue

Doctor: Hi, Emma, How are you?

Emma: I have a white discharge from my vagina.

Doctor: How long you have had   this discharge?

Emma: I think for about a week.

Doctor: Do you have any vaginal itching or soreness?

Emma: Yes, it's very itchy and I feel pain and burning when I urinate.

Doctor: Does your discharge have any odor?

Emma: No, I do not smell any odor, but it is white and thick and sticky like cheese.

Doctor: Have you been on any antibiotics recently?

Emma: Yes, I had a cold so I was given antibiotics to take for 5 days.

Doctor: How many sex partners do you have?

Emma: Just one, my husband.

Doctor: Do you have any history of STD's in the past?

Emma: No.

Doctor: Have you had any yeast infections in your vagina in the past?

Emma: No, not that I know of.

Doctor: Do you feel pain when you have sex?

Emma: No.

Doctor: How often do you do douching? Did you do it more often because of your symptoms?

Emma: Frankly speaking, yes.

Doctor: Regular douching is not a good idea when you have any vaginal symptoms. Next time, if you have a problem, come and see me, so you can get the right diagnosis and treatment. Let me examine you. Based on your history, you might have developed Candida vaginitis. It is a yeast infection that some women commonly get after taking antibiotics.

Emma: Oh, I see.

Doctor: (After exam) You have Candida vaginitis. I will prescribe an antifungal suppository to use for a week. I also did a routine PAP smear because this is a good test to pick cervical cancer early when the chance of cure is much better.

11. Carpal Tunnel Syndrome

CC: "I have a tingling sensation, numbness, and pain in my right hand."

Annie is a 30 year old female who has a tingling sensation and pain in her right hand, specifically in her thumb, index, and middle fingers for 2 months. She wakes up at night due to her pain sometimes and her pain and tingling sensations have been getting worse for the last 4-5 days. Annie had the same problem 5 years ago when she was pregnant but it resolved spontaneously, so Annie expected that these symptoms would get better like last time. After she gave birth, Annie took up piano lessons even though she was extremely busy raising her child.

However, after taking lessons, Annie noticed that her symptoms were aggravated by fine hand movements such as playing the piano, sewing, and writing - if she stopped those activities, her pain and tingling sensation would gradually get better. Annie was pretty upset because she loved these leisure time activities, but could not continue because of the pain bothering her. Recently, Annie was forced to stop playing the piano completely due to the severity and intensity of her hand pain. This made Annie depressed and it brings her to tears when talking about her symptoms, and she questioned her clinician repeatedly about when she can resume her favorite activities.

Annie came to the clinic because she noticed that she drops small objects so often due to weakened grip strength in her right hand compared to her left hand. Annie mentioned that she dropped at least 5-6 plates last month in the kitchen and she realized that it was time to seek medical help. The doctor recommended a wrist splint to help relieve her symptoms, an EMG to confirm a diagnosis of median nerve compression and a referral to a hand orthopedist for a possible carpel tunnel release procedure.

Dialogue

Doctor: Hi Annie. How are you? You don't look very happy. How can I help you?

Annie: I have pain and a tingling sensation and some numbness in my right hand and it often wakes me up at night.

Doctor: I think you had the same problem when you were pregnant. Am I right?

Annie: Yes. I had the same symptoms when I was pregnant, but my symptoms just went away after I gave birth. So, I expected it would also go away by itself. But it hasn't gotten better. Actually, I play the piano a lot, I love the sound of my piano, but then my pain gets worse.

Doctor: Anything else that you are doing that brings on these symptoms?

Annie: I like sewing and writing. If I do them for a long time, I notice my pain getting worse until I stop and rest.

Doctor: Do you still play the piano?

Annie: No, I've stopped completely. (Patient starts to cry) The pain has gotten to the point that I can't bear it anymore. It is really painful now. I love the sound of the piano and I can't imagine my life without being able to play the piano. Also, I never told you this before, but I play the violin too. However, playing the violin also brings on the same symptoms and my wrist gets very weak. So, I have dropped it without thinking twice. But piano, my piano, I want to play again. Can I do that soon?

Doctor: Annie, you are a very gifted, musical lady. Good for you. Now, please put your elbow on the desk and hold your forearm in the vertical position and flex your wrist for me. (Phalen's Test) Do you feel more pain, numbness or a tingling sensation?

Annie: Yes, this position gives me more pain.

Doctor: I think you have Carpal Tunnel Syndrome and it has recurred. I will give you wrist splint to help relieve the symptoms and I will order an EMG (electromyogram) to confirm the diagnosis of median nerve compression. Also, I will refer you to a hand orthopedist for a possible carpal tunnel release procedure if your symptoms don't improve with rest and the splint. Feel better Annie.

12. Cataract

CC: "I have trouble driving at night and I noticed my vision is blurry at times."

Eric is a 73 y/o Caucasian man who came to the clinic due to poor night vision and blurred vision for about 3 months. He lives with his 70 y/o wife Helen in suburban area and they like to go out on the weekends and Eric always drives their car when they go out. About 2 weeks ago, they went to a local restaurant for dinner and he nearly went through a stop sign because he had trouble seeing the road sign at night.

This was the second incidence. About a month ago, Eric almost hit the bumper of another car while he was driving and Helen was worried about his night vision because these two incidents both happened at night. In addition to this, he complains of painless blurred vision and light sensitivity. Eric loves reading and he wants to read almost every evening after dinner instead of watching TV, but he needed brighter lighting to read recently so they purchased an extra light stand for his reading.

Eric was diagnosed with diabetes 15 years ago and he is on insulin and he developed hypertension about 10 years ago. He is on hydrochlorothiazide and quinapril for his hypertension.

Eric's father had a cataract in the past and Eric has no knowledge of any recent eye injury. Eric has not had any previous eye surgery. Eric and Helen have been married for over 40 years and they have 3 children. All of them are married and they live in another state. They visit their parents during their vacation with their children and they are very supportive of their elderly parents.

Dialogue

Doctor: Hello, Eric and Helen how are you?

Eric: I have a problem with my vision recently. I cannot drive well at night. I nearly overran a stop sign because I could not see the road sign well. I have never had this kind of problem before and I think I need to have my vision checked.

Doctor: Do you have any other problems other than poor night vision? For example, do you have any blurred vision or light sensitivity?

Eric: Yes, my vision is blurry and when I look into the light, my eye burns.

Doctor: Do you have any pain in your eye?

Eric: No, but everything looks little bit yellow.

Doctor: Have you ever noticed you needed brighter light to read recently?

Eric: Oh, that's right. I like reading so much and almost every night I read a book after dinner. Recently, I thought one of my light bulb burned out in my study, but all bulbs were working. But I was not comfortable reading. It is like the light is too dim so I had to purchase an extra light stand just for my reading. What is wrong with my vision?

Doctor: Based on your symptoms, I think you have developed cataracts. Any of your family members have cataracts?

Eric: Yes, my father had a cataract. I remember I took him to the hospital when he needed surgery because of his cataract.

Doctor: Eric, do you have diabetes?

Eric: Yes, I am on insulin.

Doctor: Is your sugar controlled? How many times a day do you check your sugar?

Eric: I check my blood sugar twice a day before I eat. Mostly, my blood sugar is o.k.

Doctor: I am going to exam your eye to confirm the diagnosis. Eric, I will check your eyes myself though I will likely need to refer you to an ophthalmologist to more thoroughly check your eyes for cataracts, for the pressure inside your eyes (intra-ocular pressure) for glaucoma and for the back of your eye, the retina, for any disease that might be affecting your vision (age-related macular degeneration or proliferative diabetic retinopathy). He will also do vision tests. You may need medications, surgery or just a pair of glasses to correct the   problem.

The most common type of cataract is related to aging of the eye. Other causes of cataract include family history, medical problems such as diabetes, long term, and unprotected exposure to sunlight, trauma, or long- term steroid use. They usually progress gradually over a period of years.

Eric: If I have a cataract, do I need surgery?

Doctor: No, if the symptoms of cataracts are not bothering you very much, surgery may not be needed. Sometimes a simple change in your eyeglass prescription may be enough treatment.

Eric: I hope I do not need surgery.

Doctor: Let me examine your eyes then I will discuss your treatment options.

13. Chronic Kidney Disease

CC: "I always feel tired and weak."

Dena is a 75 y/o retired pharmacist who came to the clinic with complaint of persistent fatigue and weakness. Her symptoms continued for approximately 3 months at which point he proceeded to seek further medical evaluation. Her physician ordered a full evaluation and found that Dena had kidney disease. Her physician explained to her that due to the severity of her lab abnormalities he would refer her to a Nephrologist (Kidney doctor) for further intervention.

Dena and her family emigrated from Jordan at the age of 28. She retired from the pharmacy profession at the age of 67. She visits her native land once a year in order to spend time with her family and relatives.

She was diagnosed with diabetes 7 years ago and she has had a hypertension for over 20 years.

She is on irbesartan 150mg QD for her hypertension and she has a known drug allergy to ACE inhibitors. According to Dena, she develops coughing and rash if she takes an ACE inhibitor. (Angiotensin - Converting Enzyme).

She is categorized as CKD stage 5, eGFR less than 15 cc/min (estimated glomerular filtration rate, normal is 120-130cc/min). She also c/o anorexia, nausea and mild pruritus.

She did not have any history of recent infections and no skin rash was reported.

Dialogue

Dr. Kalloo: How are you, Dena?

Dena: I feel tired all the time and I feel very weak too.

Dr. Kalloo: How long you have you had these symptoms?

Dena: For about 3 months. I can't do anything because of the weakness and fatigue, so I went to see my PCD.

Dr. Kalloo: What did he say?

Dena: He did some blood tests and then he told me that my kidney disease has gotten worse and it is the time to see kidney doctor.

Dr. Kalloo: Besides your fatigue and weakness, what other symptoms do you have?

Dena: I feel nauseous and I itch all over. I have no appetite. I don't want to eat anything.

Dr. Kalloo: Did you have any infections recently? Any fevers or chills?

Dena: No.

Dr. Kalloo: Do you have any problems urinating? Do you have any pain or discomfort passing urine or are you urinating frequently? If so, are your symptoms worse at night?

Dena: No, I am O.K.

Dr. Kalloo: Any of your family members have kidney disease or kidney failure? Has anyone ever been on hemodialysis?

Dena: Yes, one of my brothers in Jordan has kidney disease.

Dr. Kalloo: I know you have diabetes. Is anyone else in your family diabetic?

Dena: My younger brother in Jordan has diabetes.

Dr. Kalloo: Do you have any abdominal pain, nauseas or vomiting?

Dena: Not that I know of.

Dr. Kalloo: You have hypertension. What blood pressure records do you get when you check you BP? I am going to check your BP shortly in here.

Dena: As far as I know, my BP is ok. I am taking only one medication for my BP.

Dr. Kalloo: Let me examine you. You have several risk factors for CKD. These include diabetes, hypertension, a family history of kidney disease as well as your age (older than 60). I will order blood and urine studies today: a CBC, basic metabolic panel (BMP), hepatic function panel, urinalysis with urine sediment, and urine albumin excretion (UAE). The amount of protein in the urine may help to determine the cause of kidney damage. Diabetic kidney disease is usually associated with heavy(nephrotic range, > 3.5 gms/ 1.73m2 BSA/ 24 hours)proteinuria while hypertensive kidney damage (hypertensive nephrosclerosis) usually has much less protein spillage into the urine. The medication you are taking, such as *ACE i and *ARB, may reduce protein loss in the urine and helps to slow down the progression of renal disease, also some people with hypertensive kidney disease may develop heavy proteinuria with very advanced renal disease. If you are anemic(hemoglobin< 10gms/dl), we will begin an ESA (erythropoietic stimulating agent) like epoietin or darbepoietin to help your body produce more blood cells and we will check your iron levels in your blood to see if you are also iron deficient, which may contribute to anemia. The ESA would not work well without adequate iron levels in your body. Anemia is a common complication of your disease and may explain the weakness and fatigue.

But some of your other symptoms are from the kidney failure itself and the accumulation of waste products (toxins) in your blood. We need to talk about renal replacement therapy as a treatment for your kidney disease. You don't have enough of your own kidney function anymore and we need to prepare you for artificial kidney treatment, either via hemodialysis through the blood or peritoneal dialysis through the abdomen. For hemodialysis, you would need a surgical procedure to connect an artery directly to a vein in your arm (an arterio-venous fistula) or a connection by an artificial vein (an arterio-venous graft). In the interim till that is ready to use, we have a central venous catheter (CVC)inserted in the internal jugular vein going through the superior vena cava to the right atrium for use for hemodialysis. For those who choose peritoneal dialysis, we have a catheter (Tenckhoff catheter) inserted in the abdomen for use. After a complete evaluation, kidney transplant may be an option for you. We do not have any time to lose and need to move quickly to get you started on some form of treatment.

*ACE i: Angiotensin Converting Inhibitor

*ARB: Angiotensin Receptor Blockers

14. Chronic Sinusitis

CC: "I have been coughing and have had a stuffy nose for about 2 months that I can't seem to shake".

Robert is a 20 y/o Asian college student who presented with a chronic cough and nasal congestion.

His symptoms started after he was helping his dad, who owns several rental properties in town, renovate one of their rental properties right after a tenant moved out. Robert was trying to get a summer job, but his dad asked him to help him instead.

According to Robert, he was sanding the wooden flooring with his dad and he breathed in copious amounts of dust while doing this work even though he wore a respirator mask. He did this work for several days until his dad finished the wooden floors completely.

He also c/o headache and nasal congestion and he became very uncomfortable because of the nasal congestion. He became easily fatigued along with these symptoms and he had to take a nap in between classes during school hours because he was so tired. He thought he had developed allergies so he purchased over the count medications at the local pharmacy, but it did not help improve his symptoms at all.

He does not have any history of asthma or other chronic respiratory problems.

Robert is a non-smoker.

Dialogue

Doctor: Hi, Robert, How are you?

Robert: I have been coughing and my nose is always stuffed up.

Doctor: When did your symptoms start?

Robert: It started about 2 months ago after I helped my dad fix some wooden flooring.

Doctor: You fixed a wooden floor with your dad? What did you do?

Robert: We had to re-sand and re-surface the whole floor and I breathed in a lot of dust while I was working with him. When I did the sanding of the floor, I kicked up a lot of dust. I wore a respirator mask to try not to breathe in so much of the dust, but I felt like I was choking on the dust after I finished the sanding.

Doctor: Do you have any other symptom other than those you have already mentioned?

Robert: Yes, I get headaches and I seem to get tired very easily. I did this job during my summer vacation and then I went back to school. During school I find I need to take a nap in between my classes because I feel so tired.

Doctor: When you cough, do you bring up any phlegm? When do you cough more? Night or day?

Robert: No. it is a dry cough and I think I cough more at night. Sometimes I can't sleep because of the coughing and it is pretty bothersome.

Doctor: Do you have any history of allergies or asthma?

Robert: I thought it was an allergy in the beginning so I bought some allergy medicine at the local pharmacy, but my symptoms never got any better. I do not have any allergies or asthma that I know of.

Doctor: Do you smoke? Have you have any recent upper respiratory infections?

Robert: I don't smoke and I haven't had any infections recently.

Doctor: Have you ever had a similar problem in the past?

Robert: No, this is my first time to have these symptoms.

Doctor: Do you have morning puffiness around your eye?

Robert: No, not that I know of.

Doctor: Let me examine you. I think you have developed an infection in your sinuses, what we call sinusitis. We will need to get some blood tests to check you white blood cell count to help me determine if you have an infection(increase in neutrophils), or if it's an allergy (increase in eosinophils). We may also need to get X-rays of the sinuses to see if there is an infection. If it looks like a sinus infection, we will try one course of antibiotic therapy. We may need to send you to an ENT (ear, nose, throat) specialist if it does not respond to antibiotic therapy for possible surgical drainage of the infection, but most people do respond to just a course of antibiotic therapy. If there is no improvement, we could test for a non- infectious cause of sinusitis such as ANCA (anti-neutrophil cytoplasmic antibody) - associated vasculitis (formerly called Wegner's granulomatosis now called microscopic polyangitis (MPA) with granulomatosis).

15. Congestive heart failure

CC: "I can't breathe well. I have shortness of breath."

Marsha is a 68 y/o Caucasian female who present to the clinic with shortness of breath upon exertion. She also noticed swelling in her ankle and leg which worsened over the past one month along with the SOB. Because of her ankle swelling, sometimes she could not wear her shoes and she have to wear sandals only. One of her hobbies is collecting shoes and she carries more than 60 pairs of shoes, but she can't even wear them because of her condition and this made her very upset She feels SOB even at rest and if she is lying down, she has breathing difficulty too. She has to use 3 pillows to sleep at night and sometimes she woke up with coughing in the middle of the night. She smoked cigarette one pack per day for 20 years then she stopped smoking 15 years ago.

Her SOB affected her daily living significantly and she can't walk even 1-2 blocks without stopping to catch her breath. She was diagnosed with hypertension 23 years ago and she is on Amlodipine 10 mg daily. Her father passed away from a heart attack at the age of 59 and her mother had hypertension. She has had chest pain episodes several times in the past, but she did not seek medical advice because it was too time consuming. She gained 7 pounds over the past 2 weeks and she had palpitations. She is on Atorvastation 10 mg for her hypercholesterolemia. She drinks on social occasions only and she lives with her husband on the first floor of the private house. She has 3 children and they are supportive of her mother. She claims she did not use illicit drugs anytime in her life. She used to take walks with her husband but she was forced to give it up due to her physical condition.

Dialogue

Doctor: Hi, Marsha, How do you feel today?

Marsha: I do not feel good. My SOB is getting worse and look at my ankles. I can't even wear my shoes. And you know me, I always love to wear shoes.

Doctor: Your shortness of breath. How does it affect your daily activities?

Marsha: I can't walk even 1-2 blocks without having to stop to catch my breath. I love to take a walk with my husband, but I can't do that anymore.

Doctor: How many pillows do you use at night to sleep comfortably?

Marsha: I have to use 3 pillows to sleep comfortably.

Doctor: Have you ever been awaken from sleep?

Marsha: Sometimes. I wake up due to coughing.

Doctor: Have you had chest pain in the past?

Marsha: Yes. I have had a chest pain several times in the past.

Doctor: Have you seen a doctor to check out your chest pain.

Marsha: No, I thought I did not need to see my doctor because they went away by themselves. I did not want to be bothered going to see my doctor.

Doctor: Have you gained weight recently?

Marsha: Yes, I have gained weight 7 pounds over the past 2 weeks.

Doctor: Do you get palpitations?

Marsha: Yes, I have palpitations at times.

Doctor: Do you have any family history of heart disease?

Marsha: Yes, My father passed away from a heart attack when he was only 59 and my mother has hypertension.

Doctor: Do you have high cholesterol?

Marsha: Yes, I am taking medication because of high cholesterol.

Doctor: Do you use illicit drugs?

Marsha: Never in my entire life.

Doctor: Do you drink or smoke?

Marsha: I drink on special occasions only and I smoked cigarette for 20 years then stopped 15 years ago.

Doctor: How is your appetite? Do you feel nauseous?

Marsha: My appetite is poor. I do not want to eat therefore, I eat very little. I feel nauseous and sometimes I vomit too.

Doctor: Do you have any abdominal pain?

Marsha: Yes, right here. (pointing at right upper quadrant)

Doctor: OK, let me examine you. You sound like your heart function is getting worse. Your body is retaining fluid that has accumulated in your feet and lungs. This is usually a sign of a weak heart, congestive heart failure, which can be a disease of the coronary arteries, a problem with the heart valves (mitral valve or stenosis or insufficiency), inflammation of the heart muscle (myocarditis), infiltration of the heart muscle (amyloidosis) or failure of the heart muscle (cardiomayopathy). We will need to get a chest x-ray and echocardiogram to evaluate the heart and heart valve function. We will start you on a diuretic regimen to help get rid of the fluid. We will also start afterload reduction therapy with an ACE (angiotensin converting enzyme) inhibitor. You will probably need a stress test to see if you have coronary artery disease. Based on the echocardiogram, after tests may be necessary (serum electrophoresis, endomyocardial biopsy). Based on the stress test, a cardiac catheterization may be necessary to evaluate the coronary arteries. If the stress test is abnormal, in the interim, we would also begin aspirin and Beta-blockers. We will also check your cholesterol as therapy for a high cholesterol may also be warranted.

16. Conjunctivitis

CC: "My right eye is red, it itches and it's constantly tearing."

Edy is a 43 year old Hispanic male who came to clinic with a red and itchy eye. Edy noticed his symptoms began 2 days ago after he went swimming in the town pool. He also complains of hyperemia, and watery discharge, but does not have ocular pain, photophobia, or blurred vision. He had no knowledge of contact with any person with "pink eye." Edy does not wear corrective lenses and can't recall any recent eye injury. Edy does not have any history of allergies and his vision is unaffected, but Edy was wiping his right eye continuously at the clinic due to a watery discharge. Edy was diagnosed with diabetes 5 years ago and he is on oral medication that controls his blood sugar well. Edy's doctor recommended that he start a regular exercise regimen.

Dialogue

Doctor: Hi, Edy, What brings you in today?

Edy: My right eye is red, itchy and constantly tearing. It is very bothersome.

Doctor: How long you have those symptoms? Did you do anything to your eye?

Edy: I went swimming to cool off because it's been very hot lately. Then 2 days later, I start to notice these symptoms.

Doctor: Do you have any other complaints other than what you have already mentioned?   Do you have any pain in your eyes?

Edy: Nope.

Doctor: How about your vision? Is it blurred or do you see double? Does light bother your eyes?

Edy: I can see clearly and the light does not bother my eyes. But this tearing is really bothering me.

Doctor: Have you had contact with anyone with pink eye recently?

Edy: No, All this happened after my trip to the swimming pool.

Doctor: Do you wear any corrective lenses?

Edy: No, I do not wear glasses. My vision is alright without them.

Doctor: Edy, I know you are itching a lot, but try not to touch your eyes with your bare hands. That could only cause more problems. It looks like you have an eye infection. We call it conjunctivitis. You probably caught it while swimming. I am going to prescribe antibiotic eye drops and it should clear up in a day or two. Let's get some blood work a fasting blood sugar and hemoglobin A1c to see how your diabetes is doing as high blood sugar could make you more susceptible to infections.

17. Deep Vein Thrombosis

CC: "I have a pain in my left leg and it is swollen."

Daisy is a 68y/o Asian female who presented to the clinic with pain in her left leg, and swelling in her ankle and foot for about 2 months. Daisy works at a grocery store where she has to sit down for prolonged periods of time cleaning, rinsing, and packing vegetables all day long. At the beginning, her pain was tolerable but the intensity of her pain gradually increased and she feels severe pain if she tries to climb stairs.

She had menopause when she was 53 and she suffers from severe insomnia, mood swings, night sweat and hot flashes. These symptoms have affected her daily life significantly and she was given hormone replacement therapy (estrogen) to ease her symptoms.

She is obese and she tried to lose weight by walking and exercising on the treadmill whenever she has time but she has not been successful so far.

She loves knitting and she does it almost all day long when she is not working.

She has no recent history of orthopedic or other major surgery or trauma.

Dialogue

Doctor: Hi, Daisy, How are you?

Daisy: I have a pain in my left leg and my ankle and foot are swollen too.

Doctor: For how long have you had these symptoms?

Daisy: I think I've had these symptoms for about the past 2 months.

Doctor: What kind of work do you do?

Daisy: I am working at the grocery store and when I am working, I have to sit down for long periods of time.

Doctor: Where does it hurt the most in your leg?

Daisy: I usually feel the pain in my calf and it is really painful now.

Doctor: Is there any change in your skin color in that area?

Daisy: Yes, it looks very red and feels warm when I touch it.

Doctor: In the past, have you ever received hormone replacement therapy after menopause?

Daisy: Yes, I suffered from severe insomnia, hot flashes, night sweats and my mood was all over the place. These symptoms affected my daily life significantly and I did not know what to do. It was a horrible feeling. I had no other choices and I was given hormone replacement therapy at that time.

Doctor: Do you smoke?

Daisy: No.

Doctor: Did you have any major surgery or an orthopedic procedure recently?

Daisy: No.

Doctor: Did you have any injury or trauma to your left leg recently?

Daisy: No. But, what is the cause of my symptoms?

Doctor: Based on your history and symptoms, I think you may have developed a blood clot in the vein of your leg (deep vein thrombophlebitis, DVT). You have several risk factors for DVT. Your age, obesity, inactivity, and estrogen hormonal therapy are contributing risk factors for DVT. Let me examine you to confirm the diagnosis.

Daisy: Doctor, what is the best treatment for me? Do I need a special test for an accurate diagnosis?

Doctor: We will get a Doppler-ultrasound immediately. If a DVT is found, you will need to be hospitalized for an intravenous blood thinner (heparin) and then you will be converted to an oral medication (Coumadin). We will also get some tests for a hypercoagulable disorder that may have contributed to the risk of forming a blood clot (deficiency of anti-thrombin III, protein C or protein S, presence of a procoagulant factor-anti-phospholipid antibodies, lupus anticoagulant, a mutation in factor V Leiden, anti-beta-2 -glycoprotein antibodies). These abnormalities would indicate the potential need for lifelong anticoagulation compared to just 6 months of therapy for an otherwise uncomplicated DVT.

18. Depression

CC: "I have lost 10 kgs over the past 3-4 months and I can't sleep well at night."

Lara is a 32 y/o (year old) single, Caucasian lady who works at the local fabric store. She came to the clinic because of continued weight loss for the last 3 - 4 months though she did not have any intention to lose weight at all, since she thinks she is already very slim compared to her height. According to her, she has lost about 10 kgs over the past 3- 4 months and she was forced to purchase a new size of clothing for almost everything. She claims that she consumes the same or greater amount of food to try to halt her unwanted loss of weight and she even tried fattening foods such as ice cream or chocolate chip cookies even though she does not like them, but she has been unsuccessful. She also complains of difficulty falling asleep and she wakes up early in the morning around 4 am, and cannot go back to sleep.

She feels very tired and can't focus on her work during the day time due to extreme fatigue from lack of proper sleep. She feels very weak physically after she lost 10 kgs and she realizes that her energy level has been decreasing too. She noticed that she has crying spells often when she is alone. Certain things that never bothered her before either at home or at work now bother her a lot and makes her cry easily. She has feelings of hopelessness and guilt almost every single day. She can't make decisions due to an inability to think and concentrate properly. She states that she wish she could hide from her family and job so she does not have to deal with anybody or anything since everything seems to bother her. She broke up with her boyfriend of 5 years about six months ago because of differences in their personalities and it was a very difficult situation for her to deal with. She thought she would be all right without him and she would meet another person, but she admitted that she still thinks about him a lot when she is alone. She denies thinking about death, suicide or hurting other people. She does not smoke, drink or use illicit drug. Her maternal grandmother was hospitalized several times due to depression when she was a teenager .Her uncle has a history of depression also, but he was never hospitalized. He is only on medications.

Dialogue

Doctor: How are you Lara? You look very tired. How can I help you today?

Lara: I keep losing weight even though I am not trying to lose weight.

Doctor: I see, so your weight loss is unintentional. How much have you lost recently?

Lara: I lost 10 kgs over the past 3-4 months and I do not know why. I am eating the same amount of food. Actually, I think I consume more food than before to try to maintain my weight because I always feel tired and very weak since I have lost weight.

Doctor: Do you sleep well?

Lara: No, I have trouble falling a sleep at night and even though I fall asleep late , I still wake up very early, like 4 o'clock in the morning and can't go back to sleep.

Doctor: Have you ever tried sleeping medications?

Lara: I thought about using an over the counter medicine, but I never actually tried one.

Doctor: How do you feel during daytime, just tired?

Lara: I can't concentrate properly about what I am doing and I have become so indecisive. It is really difficult to work all day long without getting enough sleep at night.

Doctor: Is there anything else you do or feel since your insomnia began?

Lara: I do not want to do anything. I used to enjoy reading and listening to classical music in my free time to help me relax at home, but those hobbies does not please me anymore as before. On top of that, I noticed I cry a lot when I am alone because every little thing bothers me. Most things that did not bother me before now bother me and make me cry. I have a guilty feeling and I feel like I am worthless.

Doctor: Before you started to lose weight significantly were there any major changes in your life, such as a death in your family or any problem with your relationships?

Lara: No one passed away in my family, but I broke up with my boyfriend. I have known him for years and we were discussing our engagement, but we have decided not to see each other anymore. It happened 6 months ago. I thought I would be all right, but I think about him so often, and the fact that I won't see him anymore make me so depressed and sad.

Doctor: Have you ever thought of hurting yourself or others?

Lara: No, I never thought about it. That's too scary.

Doctor: Any of your family members has a history of depression?

Lara: Yes, I can recall that my grandmother, on my mother's side, was hospitalized several times due to depression when I was teenager and my uncle also has depression and he is on medication.

Doctor: I would like to begin you on an antidepressant medication and refer you to a psychologist for therapy to help you work through these issues. Let's get some blood tests to check for anemia, sugar in the blood (diabetes), thyroid and liver or kidney disease problems. It's probably most likely related to the break-up with your boyfriend.

19. Dermatitis

CC: "My baby has a rash all over her body. It's getting worse."

Angelina is a 3 years old Hispanic girl and she presented with her mother due to a rash on her face, arm and leg. Her rash broke out one week ago on the left forearm first, then spread to her right leg and face too. Angelina scratched her rash repeatedly because of intense itching and scabs have developed but there were no signs of infection. She was very cranky for the last two days and did not sleep well at night. Actually, she cried most of the night. Her mom decided to try an over -the -counter ointment. She purchased Hydrocortisone 0.5% ointment from local pharmacy and applied it on the affected areas and she saw some degree of improvement after a couple of days, but Angelina still suffered with the rash. She was seen by her pediatrician 3 months ago during her regular well-child visit and she did not have any medical problems. She attends nursery school 5 hours a day for 5 days a week and she get along well with her playmates in the class. Sometimes Angelina visits her playmates home to play together after the class. Angelina lives with her parents and her brother who is 3 years older than Angelina. Her immunization status is up-to-date.

Dialogue

Doctor: Good morning, Mrs. Rivera, What brings you in today?

Mrs. Rivera: My daughter has a rash on her face, arm and leg and it's getting worse even though I applied ointment on the affected areas.

Doctor: Does she scratch a lot?

Mrs. Rivera: Yes, sometimes she scratches them until they bleed and I am really concerned about my daughter. After she scratched her rash a lot then she cries. It must be very painful and when it bleeds, even I feel that pain for my daughter.

Doctor: When did you first notice her rash? Where did it start first?

Mrs. Rivera: I first noticed her rash about weeks ago and it started on her left forearm then it spread over to her leg and face eventually.

Doctor: Any of your family members or her playmates have similar symptoms?

Mrs. Rivera: Not that I know of.

Doctor: Have you ever change her soaps, lotion or detergents recently?

Mrs. Rivera; No, I did not change any of these items. I have been using the same soaps, lotion and detergents for years for all my children.

Doctor: Did you take her on any trips outside of town?

Mrs. Rivera: No, she has stayed home with me after her nursery school.

Doctor: Did she have any diarrhea or vomiting along with her rash?

Mr. Rivera: No, she does not have those symptoms. What should I do at home to help her other than applying an ointment?

Doctor: You should keep her skin well hydrated, moist and avoid frequent bathing. Frequent bathing will irritate her skin more so it won't help her to recover from her skin condition.

Mrs. Rivera: But my baby loves a bath, and I love to give her a bath too. She is so adorable and cute.

Doctor: After her symptoms improved, you can give her a bath every day if you want. Also, cotton clothing and linens are the best to reduce irritation on her skin. I will send her for her skin testing to check for some type of allergies.

20. Diabetes

CC: "I have been going to the bathroom very often to urinate for the past month. Also, I get tired very easily and feel very thirsty all the time."

Alex is a 52 year old Caucasian male who visited the clinic for his annual physical exam. He recently purchased a house in a suburban area and was very busy working on his new property. The previous owner never lived on the premises and rented out the property to tenants for years, so the house needed extensive renovation. Therefore, Alex was always working on projects on his house on a regular basis. One day, he became very tired and exhausted from the strenuous work. Alex recalled he then began to get tired easily and would have to get up at night to go to the bathroom to urinate. Alex's wife insisted that he need to check in with his PCD (primary care doctor) to confirm nothing was wrong with him.

Alex stated that he had been waking up at night to urinate for the last month and then over the past week, his urinary frequency increased to 2-3 times a night, disrupting his sleeping pattern. Alex also noted that he had to urinate during the day frequently. He feels very thirsty all the time. In the beginning, Alex thought his fatigue was because of his strenuous house work, but he was still tired during days when he didn't work on the house. Alex has been working as a custodian at the local high school for the last 20 years. His father has hypertension and diabetes while his mother has had diabetes for the last 30 years. Alex's father is on oral diabetic medication while his mother is on insulin. His two brothers also have borderline hypertension.

Dialogue

Doctor: Hello, Alex, what brings in you today?

Alex: I feel very tired all the time and have to get up in the middle of the night frequently to urinate and have trouble sleeping at night.

Doctor: Do you feel pain or burning while you pass urine?

Alex: No, but it seems that I make a lot of urine.

Doctor: Do you have any fever or back pain?

Alex: No.

Doctor: Tell me about your weight history. Have you lost or gained a lot of body weight recently?

Alex: Yes, I gained about 10 kgs over the past four months, and feel terrible about this. My belly is getting bigger and bigger. I feel like that all the weight gain went to my tummy because my waist size increased four inches.

Doctor: Do any of your family members have diabetes?

Alex: Yes, both my parents are diabetic.

Doctor: Do you smoke?

Alex: Yes, I smoke about one pack of cigarettes a day.

Doctor: Have you ever considered quitting smoking?

Alex: Yes, I know it is bad for my health and tried to quit, but I wasn't successful.

Doctor: Smoking cessation is very important for everyone. Do you want to try again using a nicotine patch or gum? How about attending smoking cessation class?

Alex: I will think about it seriously this time.

Doctor: That's good. I would like you to have some blood and urine tests done to see what's wrong. You may have developed diabetes because of your weight gain and your family history, which pre-disposes you to developing diabetes.

21. Diabetic Gastroparesis (Diabetic Enteropathy)

CC: "I have constant nausea and vomiting".

Grace is a 47 y/o Hispanic female who presented with constant nausea, vomiting and bloating.

She was diagnosed with AODM (Type 2 diabetes or adult onset diabetes mellitus) 15 years ago and she developed hypertension 7 years ago. Also, she had CKD (chronic kidney disease) stage 4 estimated glomerular filtration rate or eGFR of 15 - 30 ml/min due to diabetic nephropathy.

About one year ago, patient began to have sporadic nausea and vomiting but the frequency has been increased over the past six months. These symptoms occur mostly in the morning associated with early satiety, bloating and fullness whenever the patient eats.

Patient also c/o heartburn and epigastric tenderness. Her diabetes has been poorly controlled.

Her FBS (fasting blood sugar) runs 130-160 and postprandial 2-hour blood sugar run 250 - 300 recently with HbA1c (Hemoglobin A1 C - normal is 4-6 % - a marker of the efficacy of the diabetes control) ranging from 9-11% for the past 6 months.

She stated that she checks her blood sugar 5-6 times a day to control her blood sugar better than this, but she has not been too successful. She also has hypercholesterolemia and CKD related anemia. She met with her CDE (certified diabetes educator) about one month ago to educate herself about controlling her blood sugar better, but she still has a hard time controlling her sugar at home. The patient has smoked cigarettes 1 ppd (pack per day) for the past 30 years, but she does not use ETOH (alcohol) or illicit drugs. She is on Atorvastatin (Lipitor®) 40 mg for high cholesterol and nifedipine (CCB, calcium channel blocker) 30 mg tab (tablet) ER (extended release, long lasting drug) for hypertension and glipizide 10 mg and repaglinide 0.5 mg for her diabetes.

Dialogue

Doctor: Hi, Grace, How are you?

Grace: I feel awful. I have nausea and vomiting constantly.

Doctor: How long have you had those symptoms?

Grace: Actually, it started about one year ago, I had these symptoms once in a while at first, but now they are getting worse. I have nausea and vomiting almost constantly nowadays.

Doctor: Do you have any other stomach problems such as bloating or heartburn?

Grace: Yes, I have bloating and feel full after eating just a little, mostly in the morning. I also get heartburn and I get pain in the pit of my stomach.

Doctor: How does your blood sugar runs at home?

Grace: (with deep sigh) I don't know. I do check my finger sticks 5-6 times a day, but still my blood sugars are high at home. It is so depressing. Before I eat, I get 130 - 160 and …… (pause) after my meals, it is 250 - 300.

Doctor: You spoke to diabetes educator recently. That did not help at all?

Grace: I thought I would do better controlling my sugar at home, but it's not been much different.

Doctor: Your glycosylated hemoglobin (HbA1c) is also very high because your sugar is not controlled well. The normal range of HbA1c is 4-6 %. Most diabetes expert recommends less than 6.5-7.0% for people with diabetes to try to prevent complications of diabetes, but yours is 9-11 %. We need to bring it down because this number is dangerously high. Do you have diarrhea too?

Grace: Yes, then my stomach problem is related to my blood sugar levels?

Doctor: Yes, we called it diabetic gastroparesis. This is diabetes induced autonomic neuropathy.

Grace: Autonomic neuropathy? What does that mean? Say it in plain English.

Doctor: OK. This is a nerve disorder that occurs in diabetic patients which leads to paralysis of the muscles of the stomach. You get slow or delayed gastric emptying or transit of food from the stomach to the small intestine though there is no mechanical obstruction or blockage. The food sits in the stomach and the stomach gets more and more distended till you experience nausea and then vomiting.

Grace: Then, what should I do?

Doctor: First, we need to get a special study, called a gastric emptying study. It is a nuclear medicine study that will confirm if this is the problem.

If it is normal, then you will need to see a gastroenterologist to have an endoscopy to look at the lining of the esophagus and stomach to see if you have an ulcer or gastritis, which, in some cases, can be due to an infection (Helicobacter pylori). If you do have diabetic gastroparesis, then small frequent feedings may be helpful. Also I will prescribe a medication to help with the contraction of the stomach muscle to help push the food along to the intestine to complete the process of digestion. The name of the medication is metoclopromide. It is 5 mg. Take it 60 min before meals and at bed time. I will add another oral diabetic medication called pioglitazone. Let me do glycosylated hemoglobin today to get a new baseline. When did you go to the dentist last?

Grace: More than 1 year ago. Do I have to go to dentist soon?

Doctor: Yes, regular dental exams are necessary because of the higher incidence in diabetic patients of gum disease, which may also affect diabetes control. If you have infection anywhere in your body, your sugar will also run higher. You have to come back every 2 -3 weeks for the next 2 months to check the response to therapy. If you do not get better, we can titrate the medication up. If you still don't respond to the therapy, I will refer you to a gastroenterologist to consider further   options. Grace, feel better.

22. Diabetic nephropathy

CC: "I always feel tired, weak and I do not have an appetite".

Daniel is 56 y/o AA male who came to the clinic with constant fatigue, weak and anorexia. He was diagnosed with AODM (adult onset diabetes mellitus, Type II diabetes) 15 years ago and his PCD referred him to nephrologist for the possibility of the initiation of hemodialysis.

His kidney function declined slowly and he was informed he will need hemodialysis eventually.

He felt worn out all the time and he had a swelling in his feet and legs. Because of his swelling in his feet, he could not even wear his shoes and he had to wear sandals.

In addition to this, he had trouble sleeping and insomnia made him tired throughout the day.

He also complains of nausea and vomiting.

He was diagnosed with hypertension 10 years ago and he is on an ARB (Angiotensin receptor blocker). He is on Irbesartan 150 mg QD (daily). His blood sugar was not controlled well and his Hgb A1C (hemoglobin A1c, normal 4 - 6%) is 9% - 10%. Daniel is a non-compliant patient and did not listen to his PCD or other healthcare professionals.

He did not test his blood sugar at home as instructed by his PCD just because it bothered him. Because of his poorly controlled sugar, he met with CDE (certified diabetes educator), but he stated that "I live only once and I want to eat what I want to eat and I will do what I want to do".

Daniel also has hyperlipidemia and he has proteinuria. He smokes cigarette 1 PPD.

He looks very depressed because he has to start dialysis very soon.

His e GFR (estimated glomerulo filtration rate) is 10cc/min (normal is 120-130cc/min) and he is CKD (chronic kidney disease) stage 5.

Dialogue

Doctor: How are you Daniel?

Daniel: I always feel tired, weak, and I do not have any appetite.

Doctor: How long have you had those symptoms?

Daniel: For a while, also my feet swollen a lot and my shoes do not fit any more. I can only wear sandals.

Doctor: Do you have any other symptoms?

Daniel: I feel nauseous and I vomit sometimes, plus I can't sleep well. I feel miserable.

Doctor: Did anybody discuss with you before about your declining kidney function?

Daniel: Yes, my PCD mentioned to me that I will need dialysis eventually.

Doctor: Daniel, how are your sugars running at home? Do you test regularly at home?

Daniel: ……. (Pause) I haven't been doing my finger sticks at home for a while.

Doctor: May I ask why?

Daniel: Well, I do not want to do it anymore. It seems like a big waste of my time. My blood sugar always runs high. What is the difference whether I do it or not?

Doctor: Daniel, I am sure your other doctor has discussed about your declining kidney function because of your diabetes and your other risk factors. You have to control your high blood sugar. Controlling the blood sugar is critical in your situation.

Daniel: I know that, but you live only once and I don't want so many restrictions in my life because of my disease. There are so many restrictions I have to follow from the diet and pills to exercise. When I think about all the restrictions and rules I have to follow, I feel like I am being choked to death. I am so sorry, but I do not want to live under so many restrictions. Please understand me.

Doctor: So you eat what you want and you don't want to exercise, right?

Daniel: Yes, that's right.

Doctor: Hmm…. Daniel, your kidney has been damaged a lot from your diabetes. That's why you have all of these unpleasant symptoms. Do you still smoke cigarettes?

Daniel: …… Yes.

Doctor: Do you want to quit smoking?

Daniel: I want to quit, but I don't know where to start.

Doctor: I will send you to a smoking cessation class and I will prescribe a nicotine patch to help kick your habit.

Daniel: Thanks. Doc. By the way, do I really need dialysis soon?

Doctor: Yes, the kidneys have many tiny vessels that filter waste products from your blood. High blood sugar from diabetes can destroy these vessels. Over time, the kidney is not able to do its job as well and later it may stop working completely. This is called kidney failure. Based on your blood work result, less than 15 % of your kidney function is left. At this level, you will need dialysis soon.

Daniel: What is dialysis like?

Doctor: Well, there are different kinds of dialysis, hemodialysis and peritoneal dialysis. Some types can even be done by you at home. We will get you educated on the different types to help you choose what might be best for you, but we need to get started right away. If you can be more compliant, we can eventually talk about kidney transplantation to get you off dialysis.

23. Diabetic peripheral neuropathy

CC: "I have pain, tingling and a burning sensation with numbness in my feet".

Sam is a 70 y/o Caucasian male who came to the clinic with pain, tingling and a burning sensation with numbness in his feet for about 2 months. He stated that he felt a little discomfort in his feet when he begins to walk in the beginning. Then he started to feel pain and his walking became difficult. He was diagnosed with diabetes 30 years ago and he is on Insulin. He also was diagnosed with hypertension 20 years ago. His PCD recommended that he see a podiatrist (foot specialist) at least once or twice a year for a checkup, but he has not gone to see the foot doctor for the past 2 years. About 6 month ago, he developed a foot ulcer but it did not heal well quickly even though he applied an antibiotic cream every day. His foot ulcer finally resolved a couple of months later. He is limping severely due to the pain and his pain is getting worse at night. He tests his blood sugar twice a day and his blood sugar runs 150-180 before meals and his glycosylated hemoglobin is 9-10 %. He used to smoke 1 PPD for several decades, but now he has cut back to one- half pack a day.

He is obese and his BMI (body mass index) is 33. He used to take long walks to try to lose weight before he had the problem with his feet, but now he has not been able to walk as much as before due to the pain. He wants to join a smoking cessation class to help quit smoking. Sam drinks only at social gatherings with his family and close friends.

He lives with his wife in a first floor apartment and he has three adult children.

Dialogue

Doctor: How are you Sam?

Sam: I have pain, tingling and a burning sensation in my feet. I can't walk well.

Doctor: How long have you had these symptoms?

Sam: I think it started about 2 months ago.

Doctor: Do you have any other symptoms?

Sam: Yes, I have numbness too and, because of it, walking has become very difficult for me.

Doctor: Do you visit your foot doctor regularly? When was the last time you saw your foot doctor?

Sam: I know I am supposed to go to see him at least once a year, but I haven't gone for two years in a row. I thought I would be alright without seeing my foot doctor. But then I saw my foot doctor about 6 months ago because of a foot ulcer.

Doctor: I remember it took a long time to heal the last time you had a foot ulcer. How is your blood sugar running at home? How often do you test your sugar at home?

Sam: I try to check my blood sugar twice a day and it is usually about 150 - 180 before I eat. I know that's too high.

Doctor: Sam, your glycosylated hemoglobin is 9 - 10 %. Normal range is 4 - 6 %. You have to be very careful. For diabetes patients, tight control of blood sugar is critical to try to help prevent complications from diabetes.

Sam: I know. I know…… But it is not easy. There is nothing I can eat because of my diabetes!

Doctor: Do you check your feet every day as I recommended?

Sam: I am trying, but I do not do it every day.

Doctor: You have to maintain good foot hygiene and you should avoid injury or infection to your feet. You shouldn't walk around barefoot even at home. You should wear cotton socks and comfortable well- fitting shoes. Do not wear tight socks. It can cause decreased circulation in your foot. You still smoke, right?

Sam: ……Yes, I wish I could stop smoking, but I have cut back from 1 PPD to a half PPD. At least, I am trying…..

Doctor: Your symptoms sound like you have developed diabetic peripheral neuropathy, which is a nerve disorder from your diabetes. You have several risk factors also; your weight, smoking and poorly controlled blood sugar. Diabetes is the leading cause of limb loss in the United States. When you try to walk, do you feel more pain?

Sam: Yes, I feel more pain and also I feel pain at night too.

Doctor: Let me examine you then I will order some blood work today and prescribe some medications for your pain. Glycemic control is the key to prevent worsening of this and other complications from diabetes. I will add an oral diabetes medication, Repaglinide 0.5 mg twice a day, to help better control your blood sugar. I will give you Gabapentin for your painful neuropathy. We will get arterial NIFS (non-invasive flow studies) to check the circulation in your legs and an EMG-NC (electromyography - nerve conduction) study to confirm the diagnosis of a peripheral sensori-motor neuropathy.

24. Dry eye

CC: "My left eye feels so dry and it is stinging and burning. I am constantly tearing too."

Diane is a 55 years old Caucasian woman who came to the clinic due to a feeling of dryness in her left eye. Her left eye feels very dry especially in the morning. So she tried artificial tears, but she had to use them several times a day and it became pretty bothersome to her. She also noticed a stinging and burning sensation in her left eye and is constantly tearing. Her symptoms get better in the afternoon and then she gets worse again next morning.

She has had these symptoms for about 6 months, but she did not seek medical advice sooner because she feels absolutely fine with her left eye in the afternoon. A couple of days ago, she was standing next to a smoker while she was waiting for her bus, and the smoke irritated her eye very badly and her eye began to tear so much that she then wanted to see an eye doctor.

She was diagnosed with hypertension 8 years ago and she is on diuretic for it.

Her blood pressure is controlled well with this single medication. After menopause, she has been suffering from insomnia often and she takes sleeping pills when she can't sleep after midnight.

She has worn glasses for the past 3 decades, and her vision was not affected at all in spite of these problems.

Dialogue

Dr. Kranz: Hi, Diane, How are you?

Diane: My eye always feels so dry especially in the morning. I get a stinging and burning sensation very often. It is pretty bothersome.

Dr. Kranz: Have you tried artificial tears?

Diane: Yes, I did, but I had to use it continuously, several times a day and it did not help much. My eye is constantly tearing in the morning.

Dr. Kranz: Do you have hypertension?

Diane: Yes.

Dr. Kranz: Do you take medication? If you take medication, do you know the name of your medication?

Diane: Yes, the name of my blood pressure medication is Hydrochlorthiazide. I was told it is a diuretic blood pressure medication.

Dr. Kranz: Diane, can you recall any aggravating factors for your symptoms?

Diane: I think it is closely related with my sleeping time. If I slept well the previous night, I feel much better, but if I do not get a good night sleep, I feel much worse the following day.

Dr. Kranz: Any change in your sleeping pattern?

Diane: Yes, after menopause, I have been suffering from insomnia and sometimes I have had to take sleeping pills to help fall asleep. But I try not to use it regularly if possible because I was informed it can be habit forming. Why do I have a dry eye all of a sudden? I do not like it.

Dr. Kranz: Normally, the eye constantly bathes itself in tears. By producing tears at a slow and steady rate, the eye stays moist and comfortable. Sometimes people do not produce enough tears or the appropriate quality of tears to keep their eyes healthy and comfortable. This condition is known as dry eye. The causes of dry eye are several. If somebody is on diuretics for high blood pressure, using antihistamines for allergies, sleeping pills or using pain killers, they can develop dry eyes. Also some patient develop an autoimmune disease that affects secreting glands, the SICCA syndrome, sometimes associated with Sjogren's syndrome or other disease like SLE or mixed connective disease (MCTD) In your case, you are on a diuretic for your blood pressure and you take sleeping pills occasionally. I think those two could contribute to your symptoms. Let me examine your eyes.

Diane: Thank you.

Dr.Kranz: (After the eye examination) Your tear duct is almost completely closed. That's why tears keep coming out. Also you have an infection in your eye, in the corner of your eye near the tear duct. I saw pus there.

Diane: (With her eyes brightening) Pus? In my eye? Oh, No!

Dr. Kranz: Yes. I will prescribe antibiotic eye drops. This will take care of your problem. Apply the antibiotic four times a day for 7 days. You should get better. If you do not get better, come back to me. I may need to open your tear duct.

Diane: Dr. Kranz, Are you going to open my tear duct manually?

Dr.Kranz: No. I will use an instrument.

Diane: OK. by the way, Dr. Kranz, Do you know how surprised I was when I came to see you the first time? It was a pleasant surprise. When you touched my eye lid to examine my eye, I thought a feather had touched me gently. Your touch was the softest human touch I have ever experienced in my life. At that moment, I knew you were such a good doctor. I am sure that, that kind of soft and gentle touch comes from your heart as a caring physician and I like that.

Dr. Kranz: Oh, Thank you so much Diane. You made my day.

Diane: You are welcome, but I wanted to let you know how I felt when you touched my eye lid.

Dr. Kranz: Please do not retire soon. You are not going to retire soon?

Dr. Kranz: (Smile) I have a plan after 6 years from now. I am not going to retire just yet. I still enjoy my work and my patients.

Diane: I want you to practice for long time for patients since you are an excellent and caring doctor.

Dr. Kranz: Thank you. Diane. Feel better.

25. Erectile Dysfunction

CC: "I have a problem with getting it up when I want to make love"

Jimmy is a 49 y/o Caucasian male who presented the clinic with erectile dysfunction. He is obese and he was diagnosed with diabetes 9 years ago and he developed hypertension 4 years ago when he was 45. He smokes 1 pack per day for the past 25 years and he drinks alcohol with his friend on the weekends about twice a month.

He started to notice his erectile problem over the past 6-7 months but he was not comfortable in seeking medical help and he was hoping it would get better someday.

Instead of going to see the doctor, he tried to find an answer via internet surfing and he got some information about erectile dysfunction.

He learned walking and smoking cessation are two basic steps to relieve his unspeakable problem. So he tried to walk for about 30-45 min almost every evening and tried very hard on cutting back on smoking as he had learned. Walking was no problem but he was not able to reduce his cigarette use because he had smoked for the past 30 years and is addicted to nicotine. One of his friends teased him that he could not smoke in the dormitory when he was a freshman in college and he started to smoke after his freshman year in college.

The result of this attempt to self-manage his problem was not successful. Actually, it seemed like it did not help him at all.

He was so disappointed that he decided to see his doctor even though he did not want to reveal his problem to anyone. Jimmy felt so embarrassed at the office when he told his problem to his doctor.

He said that he feels sorry for his wife even though his wife told him she understood and she is alright with it.

Jimmy said his wife is very nice to him and he wants to receive treatment for both of them.

He is on Hydrochlorthiazide 25 mg once a day, Quinapril 10 mg bid and Tenormin 25mg daily for his hypertension. Also, he is on Glimepride 4 mg daily, Pioglitazone 30 mg daily and Glucophage 500mg bid for his diabetes. He has hypercholesterolemia and he is on simvastatin 80 mg daily at bed time. Jimmy is an assistant principal of the local middle school and he had a problem recently with the superintendent of his school district and it was very stressful to Jimmy. One of his hobbies is bicycling and he enjoys bicycling a lot with his friend whenever he has some leisure time.

Dialogue

Doctor: How are you? What brings you in today?

Jimmy: ……..Hi, doctor? (Pauses) I have a problem.

Doctor: What is your problem? How can I help you?

Jimmy: Well, I did not want to tell anybody about my problem but I think I have to tell you.

Doctor: OK. You can talk. No one is in the room except us.

Jimmy: Well, (Patient talking very hesitantly) I have a problem getting an erection and I feel very embarrassed discussing it with you even if you are my doctor.

Doctor: Oh, I see. But, believe me Jimmy, you are not alone. A lot of men suffer from the same problem like you. How long have you had this problem?

Jimmy: I think I've had this problem for the past 6-7 months but I was too shy to come and discuss it with you because I was not comfortable.

Doctor: Jimmy, you have hypertension and diabetes. Do you have a current medication list with you?

Jimmy: Yes, I brought it with me.

Doctor: Let me take a look. (After review the medication list) Some of your hypertension medications can contribute to erectile dysfunction. Do you have any other health problems or conditions? I think your cholesterol is also high, right?

Jimmy: Yes.

Doctor: Your erectile problem, is it intermittent or always a problem? If it is intermittent, it could be due to a psychogenic cause like stress rather than a physical problem.

Jimmy: I think it is constant.

Doctor: Can you achieve any degree of penetration during intercourse?

Jimmy: I think my penis is not firm enough to perform penetration. When I try, I am successful maybe one out of three times.

Doctor: Have you had any changes in your sexual desire?

Jimmy: I do not think so.

Doctor: Do you have any other sexual problems?

Jimmy: No.

Doctor: Is there any problem in your relationship with your wife?

Jimmy: No, I love my wife. We love each other. Actually, I feel sorry for my wife and I want to get treatment so I can show my wife how much I love her.

Doctor: Because you have erectile dysfunction and you can't make love when you want to it, it does not mean you do not love your wife. I am sure she understands this too. Jimmy, How does your problem affect your life?

Jimmy: I am so embarrassed and I do not feel like a man anymore.

Doctor: Jimmy, What are you talking about now? Your problem is a medical condition and you need treatment. No more or no less. I am telling you. You are not alone with this problem. This is one of the most common problems for men. There are several treatment options for this problem and we are here to help you. You know smoking is bad for erectile dysfunction. Have you tried to quit smoking?

Jimmy: I've been smoking since I was a freshman in college and it is so hard to stop smoking. As a matter of fact, I got some information about my problem before I came to you and one of the solutions to my problem was to quit smoking. But I was not successful.

Doctor: Where did you get this information?

Jimmy: I surfed the internet and I got that information. I learned walking is also helpful for my problem so I walked after dinner for 30-45min when I had time. But it did not help as much as I expected.

Doctor: I am glad you came to me and you tried those two tips. Some patients try certain medications or tools on their own that they find through advertising from unreliable sources such as magazines or sex shops. The items they are selling are not really effective. They can't prove their efficacy and it is not safe to use their items without consulting your doctor.

Jimmy: Doctor, what is the cause of my erectile problem?

Doctor: Sexual arousal is a not a simple process. To have sexual arousal, several organs have to be involved such as the brain, hormones, emotion and circulation. You have several risk factors for your erectile problem. Your hypertension, diabetes, obesity, and high cholesterol are contributing factors to your problem. What is your hobby? Do you like bicycling by chance?

Jimmy: Yes, I like bicycling and I used to go out with my friends to spend time together. But I am not doing that as much as before.

Doctor: Too bad. Bicycling is a good exercise though it may not help to solve your problem. By the way, do you feel more stress now than at other times? Because stress can create this problem, too.

Jimmy: I had an argument with the superintendent of my school district because of a difference in opinion and it was very stressful for me. He is my boss but sometimes I can't stand him. I wish I could resign from my position so I do not have to deal with him. He really gets on my nerve.

Doctor: OK. I understand.

Jimmy: What is the best treatment for me?

Doctor: In many cases, erectile dysfunction is caused by chronic health problems.

For example, if your hypertension and diabetes are poorly controlled, you can have this problem. This problem can arise from:

1. Dysfunction of the autonomic nervous system such as in diabetes.

2. Circulatory problems which can be caused by atherosclerosis due to hypertension, diabetes and high cholesterol.

3. Medications, especially those used to treat hypertension, beta-blockers in particular, though almost any medication which lowers blood pressure can contribute to the problem. In your case, diabetes and blood pressure control is important. Smoking cessation will also help in the long term. We can try medications (the PDE5, phospho diesterase inhibitors) after we do a stress test to be sure your heart will not have a problem with the medication. There is also a medication that you can inject directly into the penis to help get and maintain an erection. If that does not help, there are devices such as:

1. A vacuum device applied externally to help obtain an erection.

2. Surgical implantation of an inflatable device (penile prosthesis) to help get and maintain an erection.

Jimmy: Thank you, doctor.

26. Fifth Disease (Erythema Infectiosum)

CC: "My daughter's cheeks are so red; it looks like somebody slapped her."

Yuna is an 11 y/o Asian girl who came to the clinic with her mother due to a bright red rash on her cheeks. The rash started on her cheeks and then spread to forehead over one week. Yuna's mom Helen touched her cheeks and her cheeks were warm to the touch. Helen was concerned about her daughter and checked her temperature. Yuna had a mild fever. Her temp was 99.9 F and Yuna complained of a headache also. Last Thursday, when Yuna came back from school, her rash became much more visible and red. Helen was so surprised and asked her whether anything had happened at school. According to Yuna, one of the school pranksters set off fireworks in the school cafeteria and pulled the fire alarm. The administrator of the school rushed all the students out of the school and all the students had to wait outside of the school building until everything was cleared. Yuna was standing outside for 3 hours with the other students under the sun and her rash became very intense. Yuna has a younger brother who is in same school in 3rd grade, but he does not have any similar symptoms. Yuna and her brother go to an after- school program twice a week. Yuna goes to art class and her brother goes to chess club. Both of them like their after-school program.

Dialogue

Doctor: How are you? Yuna and Helen?

Helen: Look at my daughter. Her cheeks are so red. It looks like somebody slapped her hard.

Doctor: When did you first notice this red rash?

Helen: I noticed it about one week ago, but it is getting worse.

Doctor: Where did it start first? Do you remember how it spread?

Helen: Her rash started on her cheeks first then spread to her forehead.

Doctor: Does she have any other symptoms such as fever?

Helen: Because her cheeks were so red, I felt her cheeks and they were very warm to the touch. So I checked her temperature and it was 99.9 F.

Doctor: Did she complain of headache too?

Helen: Yes, she said she had a headache.

Doctor: Her rash is very red and intense. Was she exposed to in sunlight for any prolonged period of time?

Helen: Tell me about it. She was standing under the sun for hours outside of her school last Thursday because one student set off fireworks in the school cafeteria and pulled the fire alarm. It was a disaster. All the students had to go outside until it was safe to return to the school building. When she came back from school, I was so surprised because of her face. I am pretty sure that the sun exposure made her rash worse. I was really upset at him.

Doctor: Oh, I see. Is there anybody in the family with similar symptoms?

Helen: No. everyone else is alright in my family.

Doctor: Helen, did she complain of any joint pain or stiffness?

Helen: Not that I know of.

Doctor: I think she has a fifth disease.

Helen: Fifth disease? She is in fifth grade. That sounds weird.

Doctor: The typical symptoms of this disease are a "slapped cheek" and lacy rash.

Helen: Doctor, Yuna should stay home because of the rash?

Doctor: No, Children with fifth disease may attend school because their rash is not contagious. Good hand hygiene is important it will decrease the chance of communicability (infection). Let me examine her to confirm the diagnosis.

27. Gastroenteritis

CC: "My baby is constantly vomiting and has diarrhea".

Alice is a 2 y/o Asian girl who presented with several episodes of vomiting and diarrhea since yesterday. Alice was very cranky and she cried more often than usual.

Judith, her mother, reports that Alice does not have fever, chills or respiratory distress, but Alice was rubbing her tummy when she was crying so her mother assumed she must be having abdominal pain.

Alice has a 4 y/o brother who is not sick. Both of them attend the same nursery school Monday through Friday. No other children are sick in their nursery class.

Her diarrhea is not bloody and she has voided the usual amount of urine. Her mother has had to change her diaper 5 times a day and they were soaked with diarrhea.

Judith was reluctant to feed her with her regular formula because she has been vomiting.

Judith purchased Pedialyte® (a pediatric electrolyte replacement solution) at the local pharmacy and gave it to Alice in place of her formula.

Alice is very active and she likes to play with the other children in the clinic and she clings to her mother. Alice is a well- nourished, well-developed child and she is not in any acute distress. Her skin is well - perfused, warm and moist.

She lives on first floor of an apartment building with her parents and older brother.

She has not had any major illnesses in the past.

Her immunizations are up-to-date.

Dialogue

Doctor: Hi, Judith? How is Alice?

Judith: Alice has been constantly vomiting and having diarrhea and I am really concerned about her.

Doctor: When did it start?

Judith: Since yesterday. She is very cranky and cries a lot.

Doctor: Any other symptoms such as fever, or chills?

Judith: No, I haven't noticed any.

Doctor: Did she have any respiratory distress? I mean does her breathing appear alright to you?

Judith: Yes. She is ok.

Doctor: Alice has one brother, right? How is he doing? Is he ok?

Judith: He is fine.

Doctor: Alice attends nursery. Are there any other children there sick like her?

Judith: I called her nursery school to let them know Alice will be absent today due to her illness and I asked that question. They told me everyone else is ok.

Doctor: How many times did you change her diaper? Did she wet her diaper as usual?

Judith: Yes, I changed her diaper 5 times and they were soaked with urine and the diarrhea.

Doctor: That's good, and then she is not dehydrated. Did you feed her regular formula?

Judith: I was reluctant to give her formula, so I fed her Pedialyte® until I could come to see you.

Doctor: Is she playing well?

Judith: Yes,

Doctor: Judith, you said she had diarrhea several times. Any blood in it?

Judith: No, I did not see any blood in her diarrhea.

Doctor: OK. I will examine Alice. Based on her symptoms, most likely, Alice has a viral gastroenteritis. She needs fluid to keep from getting dehydrated but she also needs electrolyte repletion. Pedialyte® was a good choice. Try to have her drink small amount of fluids (2-4 Oz) every 30-60 minutes, rather than trying to force a large amount all at one time. That might cause more vomiting. You can use a small syringe for Alice if necessary. Breast milk or formula can be continued along with the extra fluid. It should resolve by itself in a day or so. If it doesn't or she gets worse, bring her back in immediately or take her to the ER. Because she might then need to be admitted for an intravenous to rehydrate her. Keep her at home till this resolve so the other children at the nursery school are not exposed. You will need a note from me to bring to the school to let them know that it's o.k. for Alice to go back to school.

28. Gastroesophagel Reflux Disease

CC: "I have heartburn and a cough that won't go away. And I can't seem to bring anything up when I cough."

Linda is a 53 year old obese female who came to the clinic for her heartburn and regurgitation of food. She also complains of mild epigastric and retrosternal discomfort, and nausea. Patient has had these symptoms for about a month and sometimes she notes a non-productive cough, sore throat, and chest pain. When Linda developed chest pain, she became worried about her heart, and went to see her doctor to get an EKG (electrocardiogram), but no abnormality was found on it.

In addition to these symptoms, Linda experiences chronic hoarseness and a sour taste in her mouth. Linda loves fried shrimp and she consumes them at least three times a week. She noticed this would aggravate her symptoms after eating fatty meals. However, Linda's symptoms were sometimes relieved by sitting up for a couple of hours after eating.

Linda is being treated for stage 1 hypertension for the past 10 years without complications.

She drinks alcohol with her family only during the holiday season, but she is an avid smoker for the past 20 years. Linda tried nicotine patches to stop smoking, but she was not successful.

Linda is obese and her PCD recommended that she should join a health club to lose some weight with the help of a personal trainer, but did not follow her doctor's recommendation because she was very busy taking care of her elderly parents at home. Linda lives in a private house with her husband and her parents, and she is working at a publishing company. Her two children attend college in a different state and Linda has three siblings, all of whom are healthy. Linda does not have a history of GI (gastrointestinal) problems in the past.

Dialogue

Doctor: Hi, Linda, how are you?

Linda: I have heartburn and a dry cough. They bother me a lot.

Doctor: Do you have any other symptoms? How long have you had these symptoms?

Linda: I've noticed them for about a month. I have also noticed a burning sensation in my throat and chest pain sometimes. I was kind of scared when I had the chest pain.

Doctor: Did you have your chest pain checked by another doctor?

Linda: Yes, I saw my PCD and he did an EKG on me. He said it was normal. He prescribed antibiotics for my cough, but my coughing did not get any better.

Doctor: Does your heartburn get worse with either heavy meals or fatty foods?

Linda: Let me think, yes, you are right. I love fried shrimp and eat it very often. Whenever I eat fried shrimp, my symptoms seem to get worse.

Doctor: If you sit up or take antacids, do your symptoms get better?

Linda: Yes, they do. Oh, I forgot to mention, I have a sour taste in my mouth and my voice has gotten a bit hoarse lately.

Doctor: I recommended before that you join a health club and have a personal trainer to help you lose weight. What's going on with that? Are you doing anything to lose weight?

Linda: Well, as you know, I am extremely busy at work and at home, taking care of my parents. Even though I want to join a club, I do not have the time. I am serious.

Doctor: How about smoking? You still smoke, right?

Linda: Yes, smoking is another issue. My life is very stressful and when I am under tremendous stress, I need to have a smoke. I have no other choice. However, you know I tried to quit smoking by using those nicotine patches before, but I was not successful.

Doctor: Linda, I know both losing weight and cessation of smoking are not easy, but it is very important for your health because those lifestyle changes can help reduce your symptoms. You seem to be experiencing severe reflux disease with acid from the stomach coming up to the esophagus and into your throat. It can be a very serious problem in the long run, as it can damage the esophagus leading to chronic inflammation and strictures and can even get into the lung and cause damage to lung tissue.

I will tell you several tips to help avoid heartburn and other symptoms:

Try to maintain a healthy weight.

Avoid foods that can trigger heartburn such as fatty foods, fried foods, and caffeine.

Don't consume a large or heavy meal all at once.

Try to eat smaller meals.

After you eat, sit up for at least 3 hours before you lie down. Also, smoking is a big issue. If you smoke, the lower part of your esophagus will not work properly and can aggravate your symptoms by allowing stomach contents to reflux up the esophagus.

I'll also prescribe a medication to help reduce stomach acid (a PPI proton pump inhibitor or an H (histamine) - 2 blocker) which will help alleviate some of your symptoms.

29. Glaucoma

CC: "I have severe pain in my left eye, headache, nausea and vomiting all of a sudden".

James is a 65 y/o African American (AA) male who presented to the clinic with severe ocular pain, and headache along with nausea and vomiting. He also complains of blurred vision and when he looked directly into the light, he saw rainbow-colored halos around the light source. His symptoms started about 2 days ago after he had dinner with his friends, a married couple. The couple came to visit him from another state and they went to a local restaurant. They enjoyed sea food at the restaurant and came home. On that night, he could not sleep well because of nausea and vomiting so he assumed he had caught food poisoning from the sea food. He was about to go to his PCD for his nausea and vomiting the following morning, but he started to have severe pain in his left eye. He was diagnosed with diabetes 15 years ago and he wears corrective lenses for the past 4 decades due to severe nearsightedness. He used to go to see his ophthalmologist to have an annual eye checkup for his diabetes and his doctor always measures his intraocular pressure.

His ocular pressure was WNL (within normal limit) every time he had it checked by his doctor.

About 1 year ago, he suffered blunt trauma to his eye when his friend elbowed him by mistake.

After this incident, he had persistent tearing from his left eye and he had to go to see his doctor to take care of this issue. His father had diabetes for over 30 years and he passed away due to complications of diabetes.

His father had glaucoma and required eye surgery. James lives with his wife in a private house and his wife is healthy.

Dialogue

Doctor: Hello, James How are you?

James: I have pain in my left eye and a bad headache all of a sudden. On top of that, I have nausea and vomiting. It is really weird. I do not know which doctor I have to go to take care of this problem.

Doctor: How long have you had these symptoms?

James: My symptoms started just two days ago, but the eye pain is very severe.

Doctor: Do you have any difficulties with your peripheral vision?

James: Yes, I can see things well if I look straight at it, but I notice that things are blurred around the edges of my vision.

Doctor: Are your eyes sensitive to strong light?

James: Yes, when I look at the light, I see something like rainbows around the light. What is it, doctor?

Doctor: We call that "halo vision" and this is one of the symptoms of your problem. Do you have blurred vision?

James: Yes, sometimes my vision is blurry.

Doctor: Any of your family has or had glaucoma or diabetes?

James: Yes, my father died of complications of diabetes and he had glaucoma too.

Doctor: Do you have diabetes?

James: Yes, I was diagnosed with diabetes 15 years ago and I am on insulin at home.

Doctor: When was your last visit to your eye doctor? How often do you go to see your eye doctor?

James: I visit my eye doctor every year and my last visit to my eye doctor was 6 months ago.

Doctor: I am sure your eye doctor measured the pressure in your eye. Did he mention anything to you about the pressure inside your eyes?

James: Yes, he measured the pressure in my eye. He said it was normal. By the way, why do I have stomach problems along with eye problem? I had seafood with my friends the previous night and I thought I had food poisoning, but my eye pain was so severe I came to you first. Is my stomach problem related to my eye problem? I think I went to the wrong doctor.

Doctor: You are right; your stomach problem is most likely related to your eye problem. James, you have several risk factors for developing glaucoma such as your age, your ethnic background, near sightness, a family history of glaucoma and your diabetes. Based on your symptoms and your history, I think you have developed acute angle closure glaucoma. This is a medical emergency. Let me examine your eye.

30. Gout

CC: "I have horrible pain in my left foot".

Michael is a 62 years old AA (African American) obese male who presented with excruciating pain in his left big toe for the last 2 days. It was red, swollen, warm to the touch and extremely painful. He stated that he could not even bear his bed sheet brushing up against his toe. He was not able to sleep because of the pain and he had a high fever at home so he checked his body temperature, and it was 101F (38.3C). His pain started after he consumed excessive alcohol for the last three days. Recently, he lost his job as a night shift security guard and he was pretty depressed because of his unemployment status. He was looking for another job vigorously after he received the pink slip from his employer, but it is really hard to get another job, due to the depressed economy and slow job market, and this made him drink much more than his usual amount. He also consumed large amount of seafood along with his alcohol consumption. He was diagnosed with hypertension and hyperlipidemia 1 year ago and he has had diabetes for 5 years. He is on Aspirin and he also takes a diuretic, Hydrochlorthiazide 25mg, for his hypertension. He took Acetaminophen (Tylenol ®) at home for his pain, but it did not help improve his pain.

Dialogue

Doctor: How are you, Michael?

Michael: I have excruciating pain in my left foot, actually my left big toe. It's so bad, I could not sleep well.

Doctor: Did you have a fever too?

Michael: Yes, I checked my temp and it was 101 F and my left big toe is swollen, very red and tender. It even hurts if my bed sheet brushes up against my toe.

Doctor: Did you have any other symptoms?

Michael: No.

Doctor: Michael, you have hypertension. What kind of BP medication are you taking for your blood pressure? Are you on any diuretics?

Michael: I am taking Hydro something.

Doctor: Do you mean Hydrochlorothiazide?

Michael: Yes, that's right. Is it a diuretic?

Doctor: Yes, We should stop this medication immediately. Are you on Aspirin too?

Michael: Yes, I take baby Aspirin every day. My doctor recommended it.

Doctor: Ingestion of Aspirin and a diuretic can cause secondary gout, so I am going to stop those medications today. Did you do anything else, such as drinking?

Michael: To be honest with you, I drank a lot for the last three days .My employer is downsizing his business and he got rid of my position as a night shift security guard. I tried to get another job, but it is pretty difficult for me now because of the state of the economy. It seems like nobody is hiring due to the economy. I was so depressed because I am unemployed now.

Doctor: I see. How about your diet? Do you like white bread or seafood? Have you eaten them recently?

Michael: How did you know that doctor? I love seafood and white bread so I eat them a lot when I drink alone at home. Any problem with that?

Doctor: These foods have a high purine content which can cause gout attacks, so you have to avoid them. Did you take anything to relieve your pain?

Michael: I took Acetaminophen, but it did not help at all.

Doctor: Acetaminophen isn't strong enough to help to alleviate your pain. I will prescribe an NSAID, Indomethacin, for your pain. Let me check your toe and I will do some blood tests today. We will change your BP medication from HCTZ to something else that won't exacerbate your gout. Once you are better, I want to resume the baby aspirin again because it is good for your heart health, especially with your other cardiac risk factors. With these changes and if you can cut back on your alcohol use, and make the right changes in diet, your gout   should be controlled. If not, we will consider a medication to lower your uric acid level. (A Xanthine Oxidase Inhibitor)

31. Hepatitis

CC: "My wife said my eyes and skin are too yellow and I always feel tired".

Ning is a 45y/o Asian male who presented to his primary doctor with severe fatigue, icteric sclera and jaundice of his skin. He also noticed nausea, anorexia, low grade fever and dark colored urine along with these symptoms for about 1 month. Ning felt pain in right upper quadrant of his abdomen at times and because of severe fatigue, he has been unable to do his daily chores.

According to him, there was a family feud for the past several months with his siblings regarding who is responsible for caring for their elderly parents. Ning is the second son in his family and he felt he was not responsible to take care of his parents. It is a tradition in their culture that the first son is responsible for the parents, but his sister-in-law said she would not take care of their parents because they were not nice to her when she was newly wed to his brother.

Ning's parents refused nursing home placement and Ning and Ning's wife (Seol) used to go to their parents' house to take care of them after work. Ning's parents needed a great deal of help in their daily activities because they are so fragile. In the beginning, they tried their best to take care of their parents, but it became very stressful for both of them.

In addition, Ning's brother (Chen) borrowed money from Ning to pay his son's tuition and Ning obtained that money by taking a loan from his mortgage bank. Chen agreed to pay the interest on the loan until Chen could repay the loan in full, but he did not honor his word. Ning and his wife were very upset at Chen's attitude, but Chen did not even return their call. Ning and Seol became much stressed financially and emotionally because of all these family issues. Seol was upset and when she talks about Ning's symptoms, it brings her to tears.

She thought the stress on Ning caused him to develop these symptoms. Ning does not drink or smoke. Ning never used illicit drug and he does not have any history of transfusion. Nobody in his family has similar symptoms.

Dialogue

Doctor: Hello, Ning and Seol. How are you?

Ning: Doctor, I do not feel good. I feel so tired recently and Seol told me my skin and the white part of my eyes has turned yellow.

Doctor: How long have you had those symptoms?

Ning: For about one month.

Doctor: Do you have any other symptoms such as nausea or poor appetite along with these symptoms?

Ning: Yes, my appetite has been poor and I feel nauseous at times.

Doctor: Have you ever notice any fever?

Ning: Let me think, yes, Seol told me I was little bit warm to the touch a couple of times for the past 2-3 weeks, but I did not care because I did not think I had a fever.

Doctor: Do you have any belly pain?

Ning: Yes, right side of my belly, especially the top part of my belly is hurting.

Doctor: You said you always feel tired. Tell me about it in detail.

Ning: I feel like I can't do anything well. I have no energy.

Doctor: Do you smoke or drink alcohol?

Ning: Never. My father also never smoked or drank all his life. I think I took up after him.

Doctor: Have you ever notice any change in the color of your urine?

Ning: Yes, it's turned darker.

Doctor: Anyone in your family has similar symptoms?

Ning: No.

Doctor: Have you received any blood transfusions in the past?

Ning: No.

Doctor: Have you traveled back to your home country recently?

Ning: No, I was so busy caring for my elderly parents with my wife.

Doctor: Only two of you? How about your other siblings?

98

Seol: Tell me about it. My brother-in-law and his wife seem like they do not care at all. We had been taking care of them after work for several months already. It is so stressful and we are tired both physically and emotionally.

Doctor: Anything else you want to tell me?

Ning: ......Not really.

Seol: Yes, I want to tell you more about my brother -in -law. He is so mean to us.

Ning: Seol, Please do not say any more about my brother.

Seol: Why not? Your brother gave us tremendous stress by not paying back the loan we took out under our property. He knew that money was not even from our savings account. He did not even pay the interest on that loan. That's why you are sick. Because of him, something is wrong with you right now. Look at you! You are really yellow (crying). Doctor, I heard if somebody turns yellow, that person has liver disease. Is that true? My husband is so stressed right this minute because of his brother. Our hands are already full taking of our elderly parents but they do not care at all. He is the eldest son. He should take care of his parents too.

Doctor: Seol, you must be very upset at your brother-in law. I understand how you feel but please just calm down right now.

Ning: She is so sensitive about everything because of these issues. Please understand my wife doctor. I feel bad for my wife so often. It is not like she does not want to take care of my elderly parents. She is just too busy between her job and other domestic duties on top of taking care of our elderly parents.

Doctor: OK. I understand your situation is very hard to deal with right now. Are there any other siblings in your family to help you out?

Ning: I have 2 sisters, but they live far away. They come to see our parents once a month or once every other month.

Doctor: I see. Let me examine you and I will do some blood tests today to confirm the diagnosis. You might have developed hepatitis based on your symptoms and appearance. You have jaundice which is due to the accumulation of bilirubin in the blood due to either a problem with the liver processing and excreting bilirubin or a problem (blockage) in the biliary system which allows bilirubin to be drained from the liver into the intestines. Liver disease due to infections, like hepatitis A, B or C, toxicity from drugs like alcohol, an autoimmune(autoimmune hepatitis) or inherited disorder (Wilson's disease, Alpha-1- Anti-Trypsin deficiency)can cause this problem as can obstruction to the biliary system from gallstones, or other causes such as other autoimmune diseases (primary biliary cirrhosis, primary sclerosing cholangitis). We will need to get tests of your liver function and liver enzymes along with an ultrasound of the liver and biliary system. If these tests do not elucidate the problem, a biopsy of the liver may be necessary. We can start with bed rest at home and avoidance of alcohol (I know you do not drink)and acetaminophen, but if your condition worsens or we find an abnormality, for example, obstruction, or a mass in the liver, hospitalization for more testing will be necessary.

32. Herniated Disc

CC: "I have low back pain and I can't walk."

Theresa is a 49 y/o certified patient care technician (nurse's aide) who works at a community hospital in a rural area. When she works, she assists in moving patients from bed to chair.

Most of the time, she did not have any problems performing this duty, but sometimes she felt pain after she helped to move the patient. But her pain would gradually subside a couple of hours later.

One day, she felt back pain after she helped move a very obese patient even though two other co-workers helped her. Then she noted that her low back pain began to radiate down to her right leg to below the knee.

Soon thereafter, she started to limp because of pain whenever she walked. Her leg pain gradually worsened and she was not able to go to work for the past 3 weeks.

First, she tried bed rest when she was off duty, then she applied a hot pack to her lower back.

But she was in constant pain. So she went to a chiropractic doctor to try to get help to ease her pain, but it did not help either. She felt better only briefly at the doctor's office after his manipulation but her pain came right back by the time she returned home.

Theresa stated that her pain is unbearable and she can't even walk. She came to the clinic sitting in a wheel chair today and she asked for a pain killer as soon as she came in.

Her doctor prescribed oxycodone for her pain, but unfortunately she vomited all day long after she took only one pill. She could not eat or drink anything because her stomach was so upset.

She became dehydrated from constant vomiting and she had to visit the local emergency room because she felt so miserable from the constant vomiting and back pain. She got an injection of a pain killer in the emergency room since she was not able tolerate the pain killer orally.

She went to see a neurosurgeon after she visited the ER and her surgeon recommended admission to the hospital to control her pain more effectively via the parenteral route and to evaluate the possibility of surgery.

Dialogue

Doctor: How do you feel, Theresa?

Theresa: I have severe low back pain and it is getting worse.

Doctor: For how long have you been experiencing this pain?

Theresa: My pain started about 3 weeks ago.

Doctor: How did it happen? Do you remember what did you do before you started to have this pain?

Theresa: I am working at the local community hospital. One of my duties is helping patient transfer from bed to chair when necessary. From time to time, I get back pain, but it was getting better by itself until recently. This time my back pain really got me and it happened after I helped move a very obese patient with 2 other co-workers.

Doctor: Did you do anything to try to relieve your pain at home?

Theresa: I applied a hot pack and stayed in bed all day long, but I did not get better so I even went to chiropractic doctor. I felt less pain only while I was in his office while he was treating me but when I returned home, my pain came right back.

Doctor: I think you went to ER before you came to me. What was the reason?

Theresa: Tell me about it. My PCD prescribed oxycodone for my back pain and I just took only one pill. This one oxycodone tablet turned my stomach upside down literally. I vomited all day long. I could not eat or drink. When my mouth became so dry, I went to the ER. I knew I was badly dehydrated from the side effects of my pain medication. The ER doctor gave me an injection for my pain then I fell asleep. When I woke up, that ER doctor recommended that I go to see my doctor immediately.

Doctor: I see. Theresa, if you cough or sneeze, does it intensify your pain?

Theresa: Yes. Coughing or sneezing give me more pain.

Doctor: Is there any pattern to your pain? For example, is it constant or intermittent?

Theresa: My pain is constant. It is always there now.

Doctor: Do you have difficulty in urination or moving your bowels?

Theresa: No.

Doctor: Do you have any weakness in your legs?

Theresa: I do not know. I am not sure. But my pain is unbearable now. Can I have an injection for my pain right now? I can't take pain medication by mouth. I do not want to take pain killer by mouth any more.

Doctor: Theresa, you are working in the hospital. You know I do not carry that kind of injection in my office. I do not think you can go home in this condition. I will have you stay in the hospital. I will order an MRI STAT to check your lumbar spine. According to the result of the MRI, we will discuss about treatment options for you. You may have developed an acute herniation of an intervertebral disc which is now pressing on a nerve causing the pain radiating down the leg (Sciatica). Surgery may be necessary to remove the part of the disc pressing on the nerve.

33. Hypertension

CC: "I was told by my nurse friend that my blood pressure is high".

Julie is 54 y/o Hispanic female who came to the clinic for her annual physical.

She works in a factory as a manager. She went to see a nurse friend one night last week when she had a severe headache which last for 2 days .She could not go to her doctor because it was very late in the evening after work when she developed a splitting headache . Her friend Sonia, who is a nurse, checked her BP at that time and her BP was 180/100 initially. Sonia asked her to rest for 10 minutes, and then took her BP again, and it was 158/88. She claimed that she never had high BP in the past, but Sonia insisted that she needed to be checked again for her BP to see whether she had developed hypertension.

Her primary care doctor checked her BP today and it was 162/95. Her PCD recommended a low salt diet and lifestyle modification with increased exercise activity and weight loss. Her occupation leads her to a sedentary life style and she never exercises even though she registered at a local gym 3 months ago. Julie has smoked 1 and half pack of cigarettes a day for over 20 years and she drinks 4-5 cans of beers on weekends, but she denies using illicit drugs. She has been married for 30 years and has two sons, ages 28 and 26. Two of them are married already and Julie has two grandsons. Both of her sons visit Julie once or twice a month and Julie has maintained a good relationship with her sons and daughter-in laws.

Julie's father passed away because of a heart attack at the age of 65 and his 77 year old mother had a stroke 2 years ago. Her sister Nancy, who is 50 years old, is newly diagnosed with hypertension 6 months ago, and she is under the care of a doctor too.

Recently, Julie feels more stress from her job as she is in charge of her company's new project. This added stress has made Julie drink more during the weekends. A 24- hour dietary recall shows a high intake of high-sodium foods and fats, such as fast foods, prepared and canned foods.

Julie has tried to change her dietary habits to a healthier one, but she was not successful.

Dialogue

Dr. Valerie: How are you, Julie?

Julie: I came here to have my BP checked. I was told by my nurse friend that my BP was high. She checked it twice for me on the same day.

Dr. Valerie: Your friend is a nurse? Why did you go to the nurse?

Julie: I had a severe headache last Thursday while I was working. After work, I went to see Sonia who is a nurse because it was too late to see you. At that time, she checked my BP and it was 180/100. She checked it again after I rested for about 10 minutes, but it was still high at 158/88. She told me that I have to see my doctor to see whether I have high BP or not.

Dr. Valerie: What is your profession and how are things going at work?

Julie: I am a manager at a factory and recently feel under more stress from my job, after I was put in charge of a new project at my company.

Dr. Valerie: Do you smoke, drink, or use illicit drugs?

Julie: I have smoked about one and a half packs of cigarettes a day for the last 20 years, and I have a couple of drinks on weekends at social gatherings. I do not use any illegal drugs.

Dr. Valerie: Do any of your family members have hypertension, heart problems, or high cholesterol?

Julie: My father passed away when he was 65 years old from a heart attack, while my mother had a stroke when she was 75. Also, my sister was told of high BP 6 months ago, and she is on blood pressure medication.

Dr. Valerie: Julie, you have a strong family history of hypertension and cardiovascular disease. You should be very careful. You need to exercise regularly 3-4 times a week for 30 minutes at least. When you exercise, you need to get your heart pumping. You can start with walking and gradually increase the distance you walk and how briskly you walk. Eventually, running or swimming would be even better for your heart. Smoking is bad for you and you have to consider cutting down on the number of cigarettes you smoke, and stop smoking eventually. Are you interested in joining a "Smoking Cessation" class or using cessation aids like a nicotine patch or gum or pills?

Julie: I will think about it seriously this time.

Dr. Valerie: Very good, also, you should watch what you are eating. Try not to eat either high salt or high-fat foods from fast food restaurants or canned goods. It is well established that a high-sodium intake can contribute to an elevated blood pressure. Speaking of stress, consuming alcohol will not help reduce your stress level in the long run. Reduction of alcohol intake will help you reduce your high blood pressure. In place of using alcohol, find some activities you can enjoy that will help you relax even though you are extremely busy. What do you do with your leisure time?

Julie: I like reading and watching sports on TV when I have time. Those two things make me happy. Also I would like to spend more time with my grandchildren.

Dr. Valerie: That sounds good, let's do some blood work today to check your blood count, kidney function, blood sugar and cholesterol and I will do an EKG. I want you to come back every 2 weeks for the next 2 months to have my nurse check your BP. We'll talk medications if your BP remains high after you try to make changes in your diet and lifestyle. Most hypertension especially in your age range is called essential or primary, meaning we don't really know the cause. In younger people, we should look for known secondary causes which would include:

1. Pheochromocytoma - a benign tumor of the adrenal glands producing excess sympathetic hormones, epinephrine or norepinephrine, which we screen for with blood and urine test.

2. Primary aldosteronism - another benign tumor or hyperplasia (diffuse enlargement) of the Adrenal gland producing excess of aldosterone, a hormone that causes salt retention by the kidney causing hypertension, which also causes potassium loss in the urine and low Potassium levels in the blood, which we also should screen for with blood tests (Plasma renin and aldosterone levels, which would show a high aldosterone level but a low renin level).

3. Renovascular disease- renal artery stenosis usually due to atherosclerosis or fibro muscular dysplasia (focal hypertrophy of the muscle layer of the arteries) which we screen for with a renal scan or Doppler-ultrasound or MRI- though it may also require a renal   angiogram for both renal vein sampling for renin levels and an arteriogram for imaging of the arteries to definitely diagnose and potentially treat by PTA (percutaneous trans luminal angioplasty).

4. Chronic kidney disease

5. Reninoma: a benign tumor of the kidney producing excess of a hormone, renin, which cause increased production of 2 other hormones, angiotensin II which constricts blood vessels, and aldosterone, which cause salt retention by the kidney.

34. Hyperthyroidism

CC: "I am losing weight, always feel so hot and my heart is racing".

Jackie is 35y/o Caucasian female who came to the clinic due to unintentional weight loss in spite of an increased appetite. She also c/o intermittent palpitations. She has lost about 6-7 kgs over the past several months and she always feels hot. A couple of weeks ago, Jackie and her husband, Jim, were invited to their friends' son's Bar Mitzvah. The weather was cold and everyone dressed warmly, but Jackie was the only one who wore a summer dress. Even though she wore summer clothing, she still felt so hot that she had to keep drinking ice water. She even fanned herself with the program paper and had to go outside to cool off at the party. Jackie's behavior made Jim uncomfortable and Jackie and Jim had to leave early. In the car, while driving home when Jim blamed Jackie for her behavior at the party, she became very irritable. Jackie started to scream at the top of her lungs all of a sudden and broke down in tears. She could not stop crying even when Jim apologized to her. Jim could not understand her behavior. Jackie is very calm and shy and she rarely screams at anybody. She told Jim she does not know why she is very upset over petty things. Jackie also told him she can't sleep at night sometimes and that was news to Jim.

In addition to that, she had an increased frequency of bowel movements and she had to go to the bathroom several times a day. She also noticed her hair became oily compared to the past and sometimes she has a tremor in her hands. Jim recommended that Jackie go see the doctor to check out what is going on with her.

Dialogue

Doctor: Hi, Jackie, How are you?

Jackie; I am constantly losing weight and I am always feel hot, really hot.

Doctor: When did you first notice these symptoms? How much weight have you lost?

Jackie: I've lost about 6 - 7 kgs over the past several months and sometimes my heart is racing for no reason. I am scared of it.

Doctor: O.K. You do not have any history of heart disease. Right?

Jackie: No, I don't have any problem with my heart as far as I know.

Doctor: You said you feel hot. Tell me about it.

Jackie: I went to a party with my husband a couple of weeks ago, and I could not stay in the ballroom because I felt so hot. I was drinking ice water constantly and had to go in and out of the room to try to cool off. I knew some people were staring at me, but I felt like I am burning up inside. We had to leave the party early because of me and Jim was very upset with me. He was so embarrassed. I am a quiet person but I feel like my temperament has changed recently.

Doctor: Any other symptoms other than what you already mentioned?

Jackie: Yes, I have several other symptoms. I can't sleep at night and the lack of proper sleep makes me cranky during the daytime. On top of that, I lose my temper easily. When we had to leave the party early because of me, we had a big argument in the car. I did not like the tone of his voice when he questioned me on what was wrong with me so I screamed at him and then broke down and cried. Jim apologized to me finally, but I know it was not his fault. I do not know why little things set me off so easily. Also my hair has become very oily. I used to have very soft, silky hair, but my hair is not shiny anymore and I do not like that. I also notice that my hands shake sometimes for no reason.

Doctor: How about your bowel habits?

Jackie: Oh, I forgot to mention that. I used to move my bowels once a day- or once every other day, but now I have to go to the bathroom so often and it is annoying. I think I go to the bathroom at least 3-4 times a day to move my bowels.

Doctor: Any of your family members has a history of thyroid disease?

Jackie: Not that I know of.

Doctor: You may have thyroid disease. I will do your thyroid function tests along with other blood work. This can be due to an autoimmune disease of the thyroid (Graves' disease) or a benign tumor in the thyroid, causing your body to produce too much thyroid hormone which accelerates your body's metabolism. We would test for   auto-antibodies that can cause Graves' disease (thyroid-stimulating immunoglobulin, which mimic TSH-thyroid stimulating hormone or directly activates the TSH receptor).It makes you lose weight, feel hot, go to the bathroom more often and speeds up your heart. This could also give you big mood swings. We may need to do an ultrasound or thyroid scan to look for a tumor of the thyroid. We can use medications such as a beta-blockers to help control your symptoms. If we find an overactive thyroid, we can use medications like PTU (propylthiouracil) or methimazole to help control the thyroid gland. Definitive treatment may require use   of radioactive iodine to ablate part or all of the thyroid gland or surgical resection of a thyroid tumor if found.

35. Hypothyroidism

CC: "I am always cold and tired."

Betty is a 42 year old female who has been physically weak, fatigued and has noticed that she always feels cold even when everyone else feels warm for about 6 months. Betty puts on an extra blanket at night, Otherwise she cannot sleep, and has gained 7 kilograms in weight over the past several months for no apparent reason. She also c/o anhedonia (literally lost interest in everything in her daily life). She also c/o constipation for up to 7-10 days.

Betty's LMP was 2 weeks ago and she noticed her menstrual flow has been getting heavier along with these symptoms. She also noticed that her skin is very dry. She forgets things easily and her hair is also getting coarser and thinning.

Betty's husband John was recently transferred to another state for work about 3 months ago, without taking Betty or their three young children of ages 5, 9, and 11 with him. Betty has been taking care of her children all by herself for the past 3 months and finds this is a very stressful task. John is only able to come home once or twice a month due to the nature of his work.

Betty is trying to join John as soon as possible, but she has multiple things to take care of before she can move such as selling the house. Betty feels very overwhelmed between house chores and raising her children alone. Even though Betty feels constantly tired and depressed, she does not have any other health issues.

Dialogue

Doctor: Betty, you look tired today. How can I help you?

Betty: I am always extremely tired and cold. I do not know what is wrong with me. I am always so cold I can't even go to the movie theater anymore and I have to use an extra blanket at night to sleep.

Doctor: Any other problems?

Betty: I have gained about 7 kgs over the past several months and I do not know why. Also, I am constipated and I bleed heavily during my periods.

Doctor: During menstruation, how many times do you change your pad per day?

Betty: I usually change 3-4 times a day, but recently I had to change my pad 6-7 per day and I am worried about it. I think I bleed too much.

Doctor: Is there any change in your skin or hair?

Betty: Yes. My skin feels very dry and I used to have fuller hair. It was very shiny too. I used to get compliments about my hair before, but now my hair has become coarse and is thinning out. It is not as shiny as before.

Doctor: O.K. Are there any other symptoms you want to tell me about?

Betty: (with big sigh) I have lost interest in my daily life. I have to do everything since John's job relocated him after his promotion. Raising 3 young children is not an easy job for me. Just yesterday, I had to go in and out of the house at least 8 times to give them a ride and I don't get any help from my parents either because they live too far away. I am thinking about moving to where John works as soon as I can sell the house, but the bad economy is a big problem and I do not know when I can sell our property. Sometimes, my daily work load is too much to bear and I feel depressed.

Doctor: Betty, I am going to do some blood tests today to find out what is going on with you. We'll check your blood count for anemia along with some tests of kidney, liver and thyroid function as dysfunction of any of these organs can give you these same symptoms. With the thyroid, you can develop an autoimmune disease (Hashimoto's thyroiditis) where chronic inflammation can damage and reduce glandular function and produce an underactive thyroid. As soon as we find out what is wrong, I will prescribe medication to help you feel better. Thyroid hormone supplementation may be necessary.

36. Iron Deficiency Anemia (IDA)

CC: "My friends are telling me I look so pale and I always feel tired".

Janie is a 33 y/o Hispanic female who presented to the clinic with chronic fatigue and weakness for about 3 months.

She is a realtor and she is always on her feet and on the go and, sometimes she does not even have enough time to eat. She thought she was always tired because of her hectic work schedule and she does not eat well not only because of her hectic schedule but also because she has a poor appetite.

One day last week, she felt dizzy while she was with her client and she had to go home early to rest. The following day, she met with a couple of her friends. It was their monthly ladies night out. At the gathering, everyone unanimously agreed that she looked very pale and some of them told her she might be anemic and told her she should go to the doctor. Then, she recalled her episode of dizziness on the previous day.

She said she goes to a local gym to exercise, but she can't do exercise for longer than 30 min due to severe fatigue with exercise and sometimes she has dyspnea too.

She said she bleeds for 4- days using 3-4 pads a day during her menses and she does not take Aspirin or NSAID (non steroidal antiinflamatory drugs).

She is married and she has two children. They are 5 and 7 years old and she tries to spend more time with them. She also goes to her children's school to do some volunteer work in the midst of her busy schedule.

Dialogue

Doctor: Hi, Janie, How are you?

Janie: I feel very tired and I feel very weak. All of my friends tell me I look so pale.

Doctor: How long you have had these symptoms?

Janie: For about 3 months.

Doctor: Do you feel dizzy at times?

Janie: Actually, I felt dizzy 2 days ago during the day while meeting with a client and I had to go home early.

Doctor: During your menstruation, how much do you bleed? Do you bleed heavily? How many days do you bleed?

Janie: Usually, I bleed for about 4-5 days and I use 3-4 pads each day.

Doctor: Does the way you feel affect your daily activities?

Janie: Well, yes. For example, if I go to the gym to exercise, I can't do it as long as I want to, but most of my other daily activities are tolerable.

Doctor: Is there any change in the color or consistency of your stool?

Janie: No, not that I noticed.

Doctor: Do you take Aspirin or NSAIDs, such as Ibuprofen or Naprosyn regularly?

Janie: No, I use Acetaminophen when I have pain.

Doctor: Do you get headaches sometimes?

Janie: Yes. Not often, but I have occasional headaches.

Doctor: How is your appetite?

Janie: I don't have a good appetite. I am a picky eater and sometimes I don't eat if I am on the run. I am trying to change my habits to healthier ones.

Doctor: Do you like red meat or spinach?

Janie: Doc, you picked the food I try not to eat. I don't like them; therefore, I do not eat them.

Doctor: Those two foods are rich in iron. They are good for you, Janie. I will check your blood to see if you have iron deficiency anemia. Based on your symptoms, you might have IDA. IDA is the most common form of anemia especially among menstruating women.

Janie: Are there any other types of anemia?

Doctor: Yes, There are several different types of anemia, some hereditary, like sickle cell anemia and alpha and beta thalassemia, and some acquired, like hemolytic anemia, folic acid deficiency or vitamin B12 deficiency anemia , or bone marrow failure. Each type of anemia needs different treatment, but you will most likely need to get extra iron in your diet. If you really are reluctant to change your diet, I can prescribe an iron supplement for you to take once we confirm if this is the problem.

37. Juvenile Rheumatoid Arthritis (JRA)

CC: "I have pain in my joints".

Rachael is 18 y/o Caucasian female who presented with diffuse joint pains that have progressively worsened over the past few months. In the beginning, her ankle had become swollen and she had difficulty walking. Then her knee swelled up later on and was very painful. She noticed the small joints of her hands were also affected as well as her hips.

She feels very tired especially in the afternoon and she has been running a low grade fever.

She also noticed pea sized nodules that developed on her elbows. She c/o joint stiffness in the morning along with joint pains for several weeks. Her pain is a 7 out of 10 and she took Acetaminophen when she experiences severe pain. Rachael is a high school student. She enjoys her school life. She likes sports and she is on the basketball team and she also plays on the tennis team. She loved to play these sports until the pain made her quit. She plays a string instrument in the school orchestra, but the pain in her finger joints made her unable to play her instrument. She has already missed school days several times this year due to her joint pains and she feels depressed about it. She used to go out with her friend on the weekends to hang out, but she had to give up this activity also because of the pain in her knees and ankles.

She lives with her parents and 2 younger brothers and they are very supportive of Rachael.

Her parents try to stay with Rachael when she has severe pain to help her out and her two brothers try to spend time together with her as much as they can since Rachael has mild to moderate mobility problems and Rachael appreciates it a lot.

Dialogue

Doctor: How are you Rachael?

Rachael: I have pain in all my joints and it is getting worse. My ankle got swollen so much and it is so difficult to move around without pain.

Doctor: Which joints are bothering you?

Rachael: I have joint pain in my ankles, knees and fingers.

Doctor: Besides the joint pains, what other symptoms do you have?

Rachael: My joints are very stiff in the morning. After I use my hands for a while, I feel   a little bit better.

Doctor: Do you have any weight loss, fever or fatigue?

Rachael; Yes, I have a little fever especially in the afternoon and I feel tired too. But I haven't lost any weight.

Doctor: Have you ever notice any bumps on your body?

Rachael: Yes, I found a small one, it is almost the size of pea on my elbow.

Doctor: How long have you had these symptoms?

Rachael: I do not remember exactly, but for at least a couple of months.

Doctor: Your joint pains, are they on both sides?

Rachael: Yes.

Doctor: Any of your family members have rheumatoid arthritis?

Rachael: Yes, my mom and my aunt have Rheumatoid arthritis. They are in pain most of the day. Do I have the same thing, doctor?

Doctor: We do not know yet. We have to run some tests to confirm the diagnosis. Do you have eye pain or injection of the whites of your eye?

Rachael: No. My eyes are O.K. doc. (with deep sigh) I have already missed school several times this year and I was forced to quit my entire favorite after-school activities. I really want to go back to school, but I can't   because of the pain. Sometimes I feel so depressed because I had to give up almost everything I like to do in my life.

Doctor: Your pain interferes with your other daily activities significantly?

Rachael: Yes, I quit basketball, tennis and playing the cello. In addition to that, I am behind in my studies too. I have to prepare for my upcoming SAT (scholastic achievement test) to apply for college, but I can't concentrate either. The pain has taken over my life entirely.

Doctor: I see. You are a young girl and I know it is a very difficult situation for you to deal with. But cheer up. We are here to help you. I am going to give you a different pain medication. You will hopefully feel better with this medication. Also, I will do some blood tests (Rheumatoid factor/latex fixation, ANA, anti-DNA antibodies) and X-ray of the hands today to investigate your symptoms further. This may be a type of autoimmune disease like rheumatoid arthritis or lupus. Rachel, there is a support group for JRA patients and you can join that group if you want. They have a meeting every Thursday evening at 6 pm in the resource center. You do not need to call in advance. If you want, just go there.

Rachael: Thank you, doctor.

38. Leg pain

CC: "I cannot walk well because of pain in my leg and it's also feels weak."

Sue is a 35 y/o school teacher who presented with left leg pain after exercising. Her pain started after she finished her regular exercise. She worked out longer than usual because she had skipped her exercise regimen for 2 days. She usually exercises on a treadmill. She does brisk walking and jogging on the treadmill almost every day. After she finished exercising this morning, she went to the hair salon to get her hair done and she felt more pain in her leg even though she was just sitting down and waiting for her turn to get her hair done. After she came back from the hair salon, she did her usual evening chores getting ready for work the following day. Her house is a two - story home and Sue's bedroom is located on the second floor. When she tried to walk upstairs, she felt a sharp pain and her leg pain got progressively worse and worse during the evening. The pain kept her up all night and Sue has decided to see a doctor instead of going to work. According to Sue, her leg is even sensitive to a soft touch and she has weakness in her left leg compared to her right leg.

Her pain is in her anterior thigh does not radiate below the knee. Sue does not have any history of back pain or leg pain in the past.

Dialogue

Doctor: Hello, Sue. How are you?

Sue: I can't walk well because of pain in my leg.

Doctor: When and how did it start?

Sue: It started right after I finished my regular exercise on the treadmill yesterday morning. After my exercise, I went to the hair salon to get my hair done and I felt pain even while I was just sitting down to wait for my turn.

Doctor: Do you exercise regularly?

Sue: Yes, I try to exercise every day because I gained 30 pounds after my pregnancy. To be honest with you, I think I stayed too long on the treadmill yesterday. I had skipped my regular exercise routine for 2 days so I wanted to make up my exercise time. I recently purchased new clothing. They are very pretty, but they are a little tight for me right now. I want to wear them pretty soon.

Doctor: I see. Sue, where is the pain?

Sue: It is in front of my thigh.

Doctor: Anything aggravating your symptoms?

Sue: My bedroom is located upstairs in my house and when I tried to walk upstairs, I felt even more pain. My husband had to help me walk up.

Doctor: On a scale of 1 to 10, where one is no pain and ten is the most severe pain you can imagine, how would you grade your pain?

Sue: I think my pain is a 7 at least. I could not sleep because of the pain in my leg.

Doctor: Your pain is constant or intermittent?

Sue: My pain is constant. It is always there.

Doctor: Does your pain radiate to below the knee?

Sue: No.

Doctor: Do you have weakness or numbness in your leg?

Sue: Yes, I do have weakness, but no numbness.

Doctor: Do you have any problem with urination or your bowel movements?

Sue: No, I am alright.

Doctor: Do you have any history of either back pain or leg pain?

Sue: No, My back is O.K. and I never had any other problem with my leg.

Doctor: I think you might have pulled your thigh muscle while you were exercising. Your pain dose not radiate below your knee. That's good. If you feel pain behind your thigh and it was radiating below the knee, it could be something else like a pinched nerve in your back. I will give you Celecoxib (Celebrex) for your pain and Cyclobenzaprine (Flexeril) which is a muscle relaxant. Please take Omeprazole (Prilosec) to prevent stomach problems while you take Celecoxib. You can apply a heating pad to your leg. No exercising till you get better, then you can resume slowly and gradually increase your exercise regimen back to where it was previously. If it doesn't improve over the next week or so, come back to me and we may need to get a MRI to see if there is a muscle tear or any cartilage or joint injury.

39. Lupus (Systemic Lupus Erythematous)

CC: "I am always tired. I have joint pains, fever and a rash on my face".

Ruth is a 35 y/o Hispanic female who came to the clinic with chronic fatigue, joint pains with swelling and fever for about 2 month. Her fever was accompanied by severe chills and she felt very cold even during summer time. After she had these symptoms, she realized that she developed a butterfly shape rash across her cheeks and nose. She also noticed her hair was falling out and she felt shortness of breath too.

In addition to these symptoms, she has lost 40 pounds over the past 3 months and she became so worried about her symptoms. She wanted to see her doctor but she could not because she does not have access to affordable health care. Ruth is a single mother with four children through two previous marriages and she has two jobs to support her children. She was extremely busy with her two jobs and four children. Another reason she was reluctant to go to the doctor was she does not have health insurance because even though she has two jobs, they were only part time jobs and her employers did not provide health insurance. One of her friends is a doctor in her native country and sometimes she would call her to get some free medical advice over the phone. She looked depressed at the office because she is under tremendous stress from her physical weakness and responsibilities from being a single mother of four children.

She stated that she does not receive any child support from her ex-spouses. She also mentioned that when she is so worried about her health and her four children, her fingers change colors to white or blue and become painful. According to her, weight loss and the constant fever were two main reasons to seek medical advice even if she had multiple other symptoms. Her parents are upset at her because she became a single mother with four children at the age of 35 and they are not supportive of her. She has no knowledge of a family history of arthritis.

Dialogue

Doctor: How are you Ruth? How can I help you?

Ruth: I always feel tired and I have joint pains and fever too.

Doctor: How long have you had those symptoms?

Ruth: For about one month and look at my face. I have this rash across my cheeks and nose. What is this? I developed this rash after my other symptoms developed over the past month.

Doctor: That is a malar rash. Do you have any other symptoms you want to tell me?

Ruth: I lost 40 pounds over the past 3 months and I have a fever often. When I have a fever, I always feel so cold and chilly even during summer time. Actually, weight loss and frequent fevers made me visit you today.

Doctor: No other symptoms?

Ruth: My hair is falling out a lot so I have to use a scarf to cover my head. People might be thinking I have cancer and I am on chemo therapy. Sometimes I get shortness of breath too.

Doctor: Ruth, does sun exposure cause you to develop this rash?

Ruth: Let me think, yes, I think so.

Doctor: Does anything seem to trigger your symptoms?

Ruth: If I am under stress, I noticed that the color of my finger changes from white to blue and get very painful. When I am in the cold, the same thing happened. It looked weird to me. I am scared too.

Doctor: Well, everyone has stress in their lives. What kind of stress do you have in your life? Anything bothering you? Stress can cause multiple problems both physically and mentally.

Ruth: I am a single mother with four young children and I do not have a stable job. I have two part time jobs and I do not have health insurance. I know some of my decisions were not right in the past. That's why I have four children already. But having four children without a stable job is really hard to deal with. In addition to that, I am sick with no health insurance. That makes me depressed.

Doctor: I see. Your family does not support you in any way?

Ruth: I am an only child of my parents. So they expected a lot from me and sent me to a top college too but I did not get my degree because I wanted to live with my boyfriend. I married my boyfriend twice but both marriages ended with divorce. I was too young to think clearly and I did not listen to my parents even though they opposed my decision very much. My parents begged me to just finish my college education. I regret that I did not listen to my parents later on and they are still mad at me. I have a doctor friend in my home country and sometimes I call her to get some free medical advice over the phone because I do not have health insurance. Doctor, what are the causes of my symptoms?

Doctor: Based on your symptoms, I think you have an auto immune disease called lupus.

Ruth: What does that mean "autoimmune disease"?

Doctor: It means that your body's immune system, which normally functions to protect you against foreign invaders such as bacteria or viruses from outside of your body, begins to form antibodies and make immune cells that attack and damage your own normal tissues. Lupus is characterized by periods of illness, called flares, where you get sick and feel worse and then periods of wellness called remission when you feel better. I will investigate your symptoms further to confirm the diagnosis. Lupus is hard to diagnose because its signs and symptoms can mimic other diseases too.

We will get blood tests (ANA, anti-DNA, Rheumatoid factor) and X- rays of the hands to see what kind of arthritis you may have. We will also check for other organ involvement especially the bone marrow (complete blood count-white and red blood cell count and platelet count) and the kidneys (BUN, creatinine, urinalysis with urine sediment- for protein and blood in the urine).

Lupus can affect the bone marrow or cause the destruction of the blood cells in the blood stream leading to leucopenia, anemia (from marrow suppression or hemolysis) or thrombocytopenia. Also, lupus can affect the kidney causing glomerulonephritis (inflammation of the filters of the kidneys) or nephrotic syndrome (heavy protein loss from the blood into the urine).

40. Otitis Media

CC: "Pearl woke up in the middle of the night and was crying very hard while she was trying to reach her right ear."

Pearl is a 1 and half year old Asian girl who was brought in by her father. According to her dad, she was very cranky and did not sleep well at all last night. She woke up every 2-3 hours last night and cried, which is very unusual for her. Her father, Jim noticed she was trying to pull her right ear when she cries, so he assumed that it was bothering her. She was really warm to the touch. So, Jim took her temperature, which turned out to be 102 F.

Pearl had ear infections a couple of times in the past and she recovered well without any complications. She had been receiving cow's milk and she has a tendency to sleep while she is still sucking her bottle. Her parents knew this is not a good habit for her, but if they try to pull away the bottle she becomes very fussy and refuses to go back to sleep until she gets the bottle back. Jim is a smoker and has tried not to smoke when he is around her, but sometimes he has smoked inside the house because of bad weather outside.

They live in a 4-bedroom apartment with her 4 year old sister. Pearl had a runny nose for about 3 days, but she has no history of allergies. Her immunization status is up-to-date.

Dialogue

Doctor: How are you, Jim?

Jim: I am OK, but my daughter Pearl did not sleep well last night and she had a high fever. When I checked her temperature, it was 102 F. I also noticed she was trying to pull on her right ear. My daughter showed the same behavior when she had an ear infection last time. Please examine her.

Doctor: Is she voiding well?

Jim: Yes, she wet her diaper several times a day as usual.

Doctor: Has she been exposed to second-hand smoking recently?

Jim: Not that I know of. I did not smoke around her recently.

Doctor: Has she caught a cold recently?

Jim: Probably, she had a runny nose for 3 days.

Doctor: Okay, let me check her throat and ears to see if she has an ear infection. She will probably need an antibiotic and pain medication. I may prescribe a decongestant if her throat is also inflamed, to help in drainage of her middle ear through the Eustachian tube.

41. Pelvic Inflammatory Disease

CC: "I have fever and belly pain."

Santos is a 22 y/o optician who presented to the GYN (gynecology) clinic with of lower abdominal pain, nausea, vomiting and fever for 2 days. She noticed a yellow vaginal discharge along with these symptoms. She had her first sexual intercourse at the age of 18 and she has changed her partner multiples times since then. She uses an IUD (intrauterine device) for contraception and has a history of STD (sexually transmitted disease) from 6 months ago. After she was treated for her STD, she continued to have relations with the same partner only and she remained sexually active. She denied shortness of breath, chest pain, palpitations. She denied CVA tenderness (costovertebral angle) and back pain.

Most of her symptoms began within one week of her menses. When she felt warm, she checked her Temp (temperature), and it was 101F orally (38.3C) so she applied a cold compresses to her forehead and took 2 tablets of Acetaminophen, but these did not help much to reduce her high temperature.

She did not have any pain or burning sensation upon urination and no gross hematuria was reported.

Her menstrual cycles are pretty regular and she does not have any PMS (pre-menstrual syndrome)

Santos' LMP (last menstrual period) was a week ago and she was pregnant twice in the past, but her pregnancies ended up with miscarriage. She also complains of dyspareunia.

Dialogue

Doctor: Hello, Santos. How do you feel today?

Santos: I felt awful the last couple of days.

Doctor: What is wrong with you?

Santos: I have a belly pain, nausea, vomiting and I also have a very high fever.

Doctor: For how long have you had these symptoms? How high was your fever?

Santos: I've felt sick for the past 2 days. My temperature was 101 F.

Doctor: Have you ever been treated for a STD?

Santos: Yes, my doctor treated me for a STD six months ago.

Doctor: How many sex partners do you have?

Santos: I have had more than one before, but after I was treated for my STD, I have remained with the same person for the last six months. However, I do not know if he is reliable or not.

Doctor: Do you have pain or a burning sensation during urination? How about any unusual vaginal discharge?

Santos: No, I do not have any pain when I urinate, but noticed that I have developed a change in my vaginal discharge.

Doctor: Can you describe it?

Santos: Yes, it normally is white, but now the color of my vaginal discharge is yellow.

Doctor: Do you have blood in your urine? How about back or side pain?

Santos: No, I do not have any pain in my back or sides, and I do not see any blood in my urine.

Doctor: When was your LMP? Do you have any PMS?

Santos: My LMP was 10 days ago, and I do not have PMS.

Doctor: Do you have any pain when you sleep with your partner?

Santos: Yes, I feel pain when we have sex.

Doctor: Okay, I will need to do a pelvic exam to see if there is inflammation of the uterus and/or vagina. We'll do a PAP smear and a test for HPV (human papilloma virus), a pregnancy test and you may need an ultrasound to check the ovaries and fallopian tubes.

42. Peptic ulcer

CC: "I get pain in the pit of my stomach and I feel nauseous. I vomit often too."

Rose is a 47 year old Asian female who came to the clinic with gastrointestinal problems. She complains of epigastric pain, nausea, vomiting, anorexia, and she has lost 20 pounds over the past 4 months. Her pain is relieved by antacids or eating. The quality of her pain is described as nawing and burning, and on top of that, she is belching a lot which is very embarrassing for her in public and because of this, she has been trying to avoid meeting people whenever possible.

Her symptoms began 5 months ago when her husband lost a lot of money gambling at the casino.

That money was an inheritance from his father who was an owner of a prosperous high tech company back in their country. Her husband Paul is the eldest son in his family and the business was supposed to be passed on to him according to their culture, but his father hated Paul's gambling habit and did not hand down the business to him. His father chose his son-in-law over his eldest son and Paul received money instead of the business.

After Paul lost all the money from his father, he even borrowed the money from a loan shark at the casino. When Paul realized that he was not able to pay back his loan, he just ran away without telling even his wife Rose. Rose does not know Paul's whereabouts and the whole scenario made her extremely upset and some nights, she can't even sleep at all.

After she started feeling these symptoms, she is trying to ignore what happened to her and Paul, but she says that she just can't forget about it. She feels washed out and she looks pale.

She has been a smoker for over 20 years and she used to drinks at social gatherings only before, but now she drinks almost every night even though she knew drinking won't solve her issues at all.

Her father was diagnosed with gastric cancer 5 years ago, but he is doing fine now after he had surgery.

Her mother is healthy except for borderline hypertension. Her parents are very supportive of her and visit her often whenever they have time to stay with her.

Dialogue

Doctor: Hi, Rose, How are you?

Rose: I do not have an appetite and I feel nauseous. I vomit a lot too.

Doctor: Do you have pain in your stomach?

Rose: Yes, right here (pointing at her epigastric area).

Doctor: Did you do anything to relieve the pain?

Rose: I take an antacid for my pain or eat something, and then I feel better.

Doctor: When do you get this pain?

Rose: Usually at night or a couple of hours after eating.

Doctor: How would you describe your stomach pain? Is it gnawing or burning?

Rose: Both, it gnaws at me and it's burning. I am telling you, it is really painful.

Doctor: How long you have these symptoms?

Rose: (With deep sigh,) I think it started about 5 months ago after my husband lost a lot of money at the casino. That money was not our money. It was from my father-in-law. My father-in-law was wealthy because he was an owner of a high tech company in our home country. I assume my husband would inherit his father's business since Paul is the eldest son, but he is addicted to gambling and that was the end of the story. My father-in-law's business went to his son-in-law, not to my husband. That's fine. But Paul borrowed money from a loan shark at the casino after he lost all his money and he disappeared all of sudden. This made me really upset and angry.

Doctor: I see. It must be very difficult situation for you Rose. By the way, do you sleep OK at night?

Rose: No, not really. I can't sleep well because I am upset constantly and I am in pain.

Doctor: Do you do anything to help you fall asleep? Do you take any medication to help you sleep?

Rose: No, I did not take medications, but I drink. I know it is not a good choice, but what can I do? I have no other choice.

Doctor: Do you smoke?

Rose: Yes, I have smoked for the last 20 years and I smoke now more than before.

Doctor: I know it is hard to do, but cigarette smoking increases gastric acid secretion so it won't help you at all. Please try to cut back your smoking.

Rose: I know. It will not help me, but I am just too stressed right now.

Doctor: Rose, I know it is pretty tough situation for you to deal with, but hang in there. It sounds like you may have developed a stomach ulcer. We'll start you on some medication (either a H2 blocker or PPI, protein pump inhibitor) to reduce stomach acid and I would like to refer you to a gastroenterologist to consider an endoscopy to get to the bottom of this. Often, it is not due to the stress but to a special kind of bacterial infection of the stomach due to Helicobacter pylori which would require a 2 week course of 2 antibiotics (such as amoxicillin and clarithromycin) to eradicate the infection. I would also like to refer you for counseling to help with the drinking and smoking problem.

43. Plantar Fasciitis

CC: "I have a severe pain in my left sole and I can't walk well".

Amy is a 46 y/o Caucasian female who came to the clinic due to severe pain in her left sole. She felt the pain the most right after rising from the bed in the morning and she started to feel better after she walk several steps. She has had this pain for about 2 weeks. She described her pain as sharp and very tender. Her pain scale is 8 out of 10. She is an elementary school teacher and she has to stand for a long time on her feet to teach the students in the school. She is obese and her doctor recommended to her to exercise to lose weight because her BP was 152/82 one month ago during annual checkup. Her doctor asked her to come back in 2 weeks to check her BP again and it was 155/85. After she saw her doctor, she had decided to follow her doctor's recommendation and she used to walk 30-45 min after dinner every day. But her pain prevented her from brisk walking and she felt like she was putting on more weight. She is married and her husband has a hypertension. His BP is controlled well with a single medication. She has 3 children and two of them attend college in another state. Amy's mother is also obese and her mother has hypertension. Her father passed away because of a heart attack when she was a teen ager. She works as a volunteer at the local library every other Saturday. She reads the books to preschool children for 2 hours and she really likes this voluntary job.

Dialogue

Doctor: Hi, Amy, How are you?

Amy: Oh, my God, I am in horrible pain in my left sole and I can't even walk well.

Doctor: For how long have you had this pain?

Amy: I have had this pain for about 2 weeks. I was hoping it would get better without seeing you, but I was wrong. It is getting worse.

Doctor: Where does it hurt the most? Show me the exact spot.

Amy: Right here (pointing to the arch of her foot)

Doctor: Do you feel more pain with your first steps in the morning?

Amy: Yes.

Doctor: Do you feel better after you walk a few steps?

Amy: Yes, but it is really bothering me a lot. My doctor recommended walking 3-4 times a week at least to help control my BP, but because of this pain I can't walk well.

Doctor: Please stand on your toes. Do you feel more pain?

Amy: Yes, I can't stand on my toes because of the pain.

Doctor: Did you do anything to try to relieve your pain?

Amy: Not really. I did not know what to do.

Doctor: Actually, the best treatment for your condition is……

Amy: Losing weight?

Doctor: You said it, but thanks for saying it. If you lose weight, the pressure on your foot will decrease and you will feel a lot better. Obesity is one of the most predisposing factors to this condition.

Amy: I will try to lose weight, but other than that, what should I do to relieve my pain?

Doctor: I will tell you a couple of tips to relieve your pain. You should rest your foot first then do stretching, flexibility, and strengthening exercises. Also, over-the-counter arch supports may be beneficial and an ice pack often relieves the pain. Avoid flat shoes when walking and barefoot walking. For some patients, simply changing the shoes is all they need. Shoes that are too small or worn often aggravates the pain. Custom orthotics and heel cups are helpful to some patients so I would try that too. If you have no improvement within 2 weeks, come back to me you may need Cortisone injections to make you feel better.

Amy: Oh, No, I do not want to get an injection. That must be very painful.

Doctor: I hope you do not need that injection. Feel better Amy.

44. Pneumonia

CC: I have been coughing for 5 days and I bring up greenish yellow phlegm."

Jane is a 67 year old Caucasian female who presented with a productive cough. She also complains of fever, chills, sweats, chest discomfort, mild shortness of breath and general malaise. Her symptoms started after she spent a long time at the new mall that opened up 10 days ago. After Jane came back from long hours of shopping, she was very tired and exhausted that day. Jane developed fever and chills soon afterwards. She recalled she felt chilly during shopping because she wore only a thin sweater instead of a thick jacket. Jane took 2 tablets of Tylenol every 6 hours for her fever and then went to bed very early to get some rest.

The following day, she still felt general malaise and mild SOB (shortness of breath). Jane stayed in bed all day long and she took 2 tablets of Tylenol every 6 hours for her fever.

Her coughing persisted, so Jane purchased an over-the-counter cough medication. But Jane's coughing did not improve and she began to bring up copious amounts of greenish yellow phlegm.

Jane received her annual influenza vaccine this year but she denied receiving the pneumococcal vaccine in the past 5 years. She was diagnosed with stage 1 hypertension 20 years ago, but her BP is well controlled on medications without any complications. Jane is a retired teacher's aide and has 3 adult children. She teaches English at the community center after her retirement as a volunteer, and she is very satisfied with her new career.

All of Jane's children are married and are in good health, except for a son with hypertension.

Her son has been on medication for 3-4 years, and his blood pressure is well-controlled. Jane's husband died of a heart attack at the age of 71. She does not have any history of tuberculosis, diabetes, or cancer.

Dialogue

Doctor: How are you feeling today, Jane?

Jane: I have been having fever and cough with a lot of phlegm for the last five days. I bought cough medication at the local pharmacy, but it hasn't gotten any better.

Doctor: What is the color of your phlegm?

Jane: It is greenish-yellow.

Doctor: Do you have any chest pain when you cough?

Jane: Yes, I get pain the middle of my chest when I cough, and I also feel very tired and exhausted when I cough.

Doctor: You said all this started 5 days ago?

Jane: Yes, I went shopping at the mall but inside it was so hot and crowded that I took off my jacket and only wore a light sweater for most of the day. I began to feel feverish and had some chills later during the day.

Doctor: Did you receive the pneumonia vaccine? How about your influenza vaccine for this year?

Jane: I got the flu shot already for this year, but I didn't want the pneumonia vaccine.

Doctor: Is there anyone you've come in contact with recently that had respiratory symptoms also?

Jane: Let me think, oh yes. One of my students was coughing during class about 1 week ago. She did not attend class for the last 4 days, so I assume she has been sick at home.

Doctor: All right, let's have you go for a chest x-ray today and some blood work to see if you have pneumonia. If the chest x-ray shows you have it, you will need treatment for "walking" pneumonia or maybe hospitalization depending on how bad it looks, I mean how extensive the infection has spread. We can do urine tests to look for antigens of pneumococcus and Legionella as an indirect way to diagnose the cause of pneumonia though we would like to also get sputum for examination (gram stain) and cultures. It the X-ray shows an extensive pneumonia, we will check your blood oxygenation saturation with an oxygen saturation monitor (pulse Ox) to see if hospitalization is required. Once you are over this, you should get the pneumococcal vaccine. It is recommended for all adults over the age of 50 once every 5 years.

45. Polycystic Kidney disease

CC: "My Doctor told me my kidneys have a serious problem".

Rosen is a 41 y/o Caucasian female who presented to the clinic with weakness, fatigue, nausea and vomiting. She always feels so tired and she looked pale.

She visited her PCD annually but she was told she does not have any problems except hypertension. She was diagnosed with hypertension 3 years ago and she is taking quinapril.

She caught a cold about 1 month ago and she took some over -the -counter medications for her cold. She felt better after a couple of days then she started to feel very lousy in general.

She decided to see her doctor and her doctor did some blood tests. After she saw her doctor, she started to feel nauseous and she even vomited while she was waiting for the results of her blood works. Two days later, her doctor called with severe abnormal results of her blood work. Her creatinine was 21 mg/dl and her BUN was 152 mg/dl. She did not understand the meaning of the results of her blood work. Her doctor told her that she has serious kidney failure and she needed to see a nephrologist immediately.

She went to see a nephrologist who did some additional tests including a sonogram of her kidneys. When she met the nephrologist the second time she was told her kidneys were full of cysts and she would need either a kidney transplant or dialysis. That was news that came out of the blue to Rosen.

Rosen has a 17 y/o son who is a 6 foot 2 inch tall football player in high school.

One day, he was hit by a football in his right flank very hard, then he developed gross hematuria and was rushed to the ER close to the school. Rosen's son Albert was diagnosed with the same disease just like his mom, polycystic kidney disease.

This happened while she was under the care of the nephrologist and Rosen is very upset that Albert has the same disease like her.

She has no previous known family history of kidney disease but both of her parents have hypertension. Rosen has been working in the cafeteria at the local elementary school over the past 12 years.

Dialogue

Doctor: Rosen, How do you feel today?

Rosen: I feel weak, tired, and nauseous and sometimes I vomit too. So I went to see my PCD and he did some blood tests. Right after he received the result of my blood work, he told me that I have to see a kidney specialist immediately.

Doctor: Do you have your blood test results with you?

Rosen: Yes, I brought them with me.

Doctor : (after reviewed Rosen's blood work result), your kidney function is extremely abnormal. I will order a kidney sonogram today and come and see me again immediately after I get the result of your test. This will need immediate attention and probable hospitalization. Do you have any other medical problems?

Rosen: I was diagnosed with hypertension 3 years ago.

Doctor: Do you take medications for your high blood pressure?

Rosen: Yes, I am taking Quinapril 10mg once a day.

Doctor: For now, stop the quinapril.

(Rosen visited doctor again the next day.)

Doctor: Rosen, do you have any back pain or flank pain?

Rosen: Let me think, yes, once in a while, I get flank pain. I thought it happened because I was so tired. My pain was not constant so I thought it was nothing serious.

Doctor: Have you ever noticed distention of your abdomen as you've gotten older?

Rosen: I have been obese all my life so I do not know.

Doctor: Have you ever had blood in your urine or frequent urination?

Rosen: No.

Doctor: Rosen, Do you have any family history of kidney disease?

Rosen: Not that I know of. Doctor, what is the cause of my symptoms?

Doctor: Your kidneys have failed. When did you first notice your symptoms?

Rosen: What does that mean my kidneys have failed? I have felt this way for the past 3 weeks only. I thought my kidney problem would get better if you treat me. My condition is not temporary? I don't get it. Doctor, please tell me about it in detail.

Doctor: ......Rosen, I am very sorry to tell you this. Your kidneys are full of cysts and the cysts are noncancerous (benign) sacs that contain fluid like water. Your condition is called polycystic kidney disease and you need either dialysis or kidney transplantation to prolong your life. Here though, we can't wait because of your symptoms and we must start dialysis immediately.

(Rosen came back to her nephrologist later then told her doctor about her son Albert.)

Rosen: Doctor, I can't believe this. My son is in high school and he is a football player. Last week, he was hit by a football on his right flank very hard then he had to go to the ER due to passing blood in his urine. The ER doctors did some tests then told us his kidneys had a lot of cysts. The same thing you said about me and he has polycystic kidney disease. Oh, my God, poor thing. What should I do with my son? I have it and my son has it. Does it run in my family? My parents did not have any kidney problems as far as I know.

Doctor: You and your son have PKD (polycystic kidney disease) caused by a mutation in a gene that is part of environmental sensing organelle of the cells of the kidney and when abnormal leads to uncontrolled growth of these kidney cells and the development of cysts which distorts the normal kidney architecture and causes the kidneys to malfunction. It is usually inherited and your son got it from you. You either got it from one of your parents or developed it de novo (a germline mutation). In some patients with PKD, they have no known family history of the disease and even though they are affected by PKD, there is a possibility that they do not know they have PKD because they do not show signs or symptoms of PKD.

There are two main genes associated with ADPKD (autosomal dominant polycystic kidney disease).PKD 1 accounts for 85 % of all cases and PKD 2 for about 15 %.

It is an autosomal dominant disease- meaning you need only one copy of the abnormal gene to get the disease and the genes are not on the sex X or Y chromosomes so that males and females are equally affected. Thus, inheritance is not linked to gender.

Rosen: That's why I have it and my son got it?

Doctor: Yes. The disease is referred as a 2-hit model- you inherit 1 abnormal gene and then something turns off the other gene. It means that not everyone who inherits the abnormal gene will definitely get the disease, but most will.

The severity or degree of abnormality of the defect in the gene determines how soon someone will develop ESRD (end stage renal disease) - but it tends to occur around the same age within the same family. ARPKD (autosomal recessive polycystic kidney disease) is due to a different gene defect. Recessive means you need both copies (one from the mother and one from father) to be abnormal to develop the disease.

Rosen: What's going to happen to my son? He is only a teenager. What if he needs dialysis like me? I gave him bad genes. I feel bad.

Doctor: You are symptomatic now from the kidney disease and must start dialysis immediately. We need to get a central venous catheter placed in the internal jugular vein passing into the superior vena cava so that we can begin hemodialysis immediately. After a few sessions, I am sure you will feel a lot better. Then we can talk about your options; an arteriovenous fistula or graft in the arm to remain on hemodialysis or a Tenckhoff catheter in the belly to change to peritoneal dialysis. While doing this, we can refer you for evaluation for a kidney transplant to get you off dialysis. We do not have good treatment now to prevent PKD from progressing to ESRD. I am afraid your son may suffer the same fate over the next 10-15 years but maybe newer treatments being tested now will be found to help slow the disease progression. But in the interim, he needs close medical follow-up and good BP control if he develops hypertension. He needs to know his children have a 50-50 chance of inheriting the disease. Rosen, do you have other children?

Rosen: I have a daughter who is 4 years older than Albert. Do I have to bring her to you too?

Doctor: Yes, she should be screened for the disease by an ultrasound of the kidney so that she can be prepared for any eventuality and get genetic counseling if she wants to have children in the future.

Rosen: What are the benefits of having genetic counseling for my daughter?

Doctor: The purpose of genetic counseling is to determine if she carries the genetic defect and the risk of passing it in to her children. Pre-natal testing of the fetus is possible if your daughter wishes during pregnancy.

Rosen: Thank you. Doctor.

46. Herpes Zoster (Shingles)

CC: "I have a painful rash under my left breast and belly".

Donna is a 67 y/o AA (African American) female who presented with a rash under her left breast and abdomen which was very painful and itchy. She noticed this rash about one week ago. She did not have fever initially, but then noticed fever and fatigue later on.

She has developed a headache, malaise and photophobia after the eruption of the rashes.

She was diagnosed with multiple myeloma (a type of cancer of the blood) 2 months ago and she is on chemotherapy. She is on Bortezomib and Dexamethasone for the multiple myeloma.

She has had hypertension for over 20 years and she is on multiple medications for it.

She also has hyperlipidemia and she has a history of a CVA (cerebrovascular accident) 5 years ago.

She is allergic to Penicillin to which she developed hives. She worked in the school cafeteria before she retired and she goes to a local nursing home 2 times a week to do a volunteer work.

She lives in an apartment with her daughter and her daughter is very supportive of Donna.

Donna has two sons who live in another state. She attends church every Sunday and she likes to attend fellowship meetings after the service with church friends.

Dialogue

Doctor: Hello, Donna, What brings you in today?

Donna: I have a rash under my breast and on my belly. It is very painful and itchy.

Doctor: When did you first notice the rash?

Donna: About one week ago.

Doctor: Your rashes, does it look like little water bubbles? Are they in groups or did they come out separately?

Donna: Yes, they look like little bubbles and they seemed to come out all at once.

Doctor: Do you have any other symptoms?

Donna: I had a slight headache, and I feel very tired.

Doctor: Do you have any pain around your eyes?

Donna: No.

Doctor: If you look directly into the light, does the light bother you?

Donna: Yes, it burns my eye.

Doctor: Have you ever had chickenpox?

Donna: Yes, I had chickenpox when I was a child. My mom told me.

Doctor: Are you on any medications?

Donna: Yes, I was just diagnosed with multiple myeloma about 2 months ago. My doctor put me on two medications.

Doctor: Do you have your medication list with you?

Donna: Yes, my daughter told me to always carry my medication list with me and give it to the doctor. Here is my medication list. You can take a look.

Doctor: That's good. You have a very devoted daughter. (After checking the medication list) Yes, you are on special medications that can suppress your immune system. Let me examine you.

Doctor: (After exam), Donna you have shingles. I think you developed this rash because your immune system was weakened because of the medications. I will prescribe an oral anti-viral antibiotic medication called acyclovir. You have to take this medication   5 times a day for about 5 - 7 days.

Donna: Five times a day? Oh, my God. How can I take the medication five times a day? I already take so many medications because of my high blood pressure and high cholesterol. Can I take a bigger dose and take it only 3 times a day like other   medications?

Doctor: (Smiling) Donna, That sounds very smart, but this particular medication would not work as well unless you take it 5 times a day. This is a short acting medication. I know it is not easy to take medication 5 times a day, but it is very important and it's only for a short period of time.

Donna: If this is the reason, I guess I can take it 5 times a day.

Doctor: I want to do some blood tests, but I am also very concerned about your other symptoms and we need to have you see an eye doctor immediately. If the infection is also in your eyes, you will need to be hospitalized immediately to get the same antibiotic through the vein. Also you need to stay away from babies, pregnant women and anybody who is on medication that suppresses their immune system until your rash heals and is dried completely. Your lesions are potentially infectious when the bubbles form and burst and can be easily spread to other people who are susceptible.

47. Subconjunctival Hemorrhage

CC: "I have redness in my right eye."

David is 57 years old Asian man who presented to the clinic with redness of his right eye.

He noticed this redness in his right eye after dinner and he was pretty scared of it because it looked like his eye bled without his knowledge. He did not have eye pain and no discharge was reported from his eye. His vision was not affected.

He was diagnosed with diabetes about 10 years ago and with hypertension 5 years ago. He is on Metformin, Aspirin, and Quinapril for his chronic conditions.

He lives in a suburban area and he likes to take care of his garden and he moved a very heavy plant on his own that afternoon while he was gardening. He stated that it was really heavy and, in retrospect, he should have waited for his son to come home so they could move it together, but he could not wait for his son. Then he developed this redness in his right eye.

Even if his eye looked scary due to the redness, he did not have any other symptoms so he came to the clinic the following day instead of going to emergency room.

Dialogue

Doctor: Hi, David. What brings you in today?

David: Doctor, look at my eye. Do you see the red spot in my right eye? What is it? I am so scared of it.

Doctor: Do you have pain in your eye? How about your vision?

David: I do not have pain in my eye and my vision is ok.

Doctor: When did you notice this redness in your eye?

David: I saw this redness in my eye last evening after dinner.

Doctor: Did you do anything strenuous yesterday such as lifting any heavy object? How about a coughing or sneezing fit?

David: I did move an extremely heavy plant on my own in the garden yesterday. Actually I brought it inside the house because my wife loves that plant and she wanted to keep it inside, otherwise, that plant would not survive in cold weather.

Doctor: (Smile) David, you must be a good husband. By the way, are you taking Aspirin?

David: Yes, I am taking a baby Aspirin every day because my doctor recommended it.

Doctor: Is there any discharge from your eye?

David: No.

Doctor: Have you ever had similar problem in the past?

David: No, This is my first time. I never had redness in my eye before.

Doctor: David, do you have any sensation of something in your eye?

David: No.

Doctor: I will measure your visual acuity. (After exam) Your eye is all right even though it looks scary, as you said. The redness in your eye is a small hemorrhage in the white of your eye. It may be result of trauma, a bleeding disorder or a sudden increase in venous pressure such as from vigorous coughing. I think, in your case, when you lifted that heavy plant, the pressure inside your eye increased suddenly and then the vessel in your eye ruptured. Especially, if one is on Aspirin, they have a tendency to bleed easily.

David: Doctor, What are you going to do with the bleeding in my eye?

Doctor: That will fade over the next couple of days to yellow then disappear, like when you   get a bruise on your body. So you do not need treatment to get rid of this blood. That blood will clear in a 1-2 weeks period. If you develop impaired vision, eye pain, or a foreign body sensation in your eye, go to your eye doctor for further evaluation, but I think you will be OK.

48. Tennis elbow

CC: "My arm is on the fire and I can't sleep due to the pain in my elbow".

Louis is a 45 y/o Caucasian man who presented with severe pain in his right elbow for about 3 weeks.

He could not move his arm freely due to the pain and he could not even raise his arm above his head. He also noticed swelling over his right elbow along with tenderness. His profession is a plumber and his occupation required using a wrench and screwdrivers many times a day.

He wanted to see a doctor earlier, but his hectic work schedule prevented him from visiting his doctor sooner. Due to the nature of his profession, he is busier during winter time than other seasons and sometimes, he has had to work until 10 pm to respond to all the emergency calls he receives.

He took Acetaminophen for his pain and wrapped his elbow with an elastic bandage, but it did not help much to relieve his constant pain. He sees his PCD annually and he does not have any health problems.

Dialogue

Doctor: Hi, Louis, What brings you in today?

Louis: My arm is on the fire. Oh my God. I can't even sleep at night because of the pain.

Doctor: For how long have you had this?

Louis: I have had the pain for about 3 weeks. But I could not come in earlier because of my work schedule. Winter is the busiest season for me, doc.

Doctor: What is your profession? Do you use your hands a lot?

Louis: I am a plumber and I work with my hands all day. I have to use wrenches and screwdrivers to fix people's plumbing and pipe problems.

Doctor: Do you have to work long hours?

Louis: Yes, sometimes I don't finish my work till 9 or 10 pm. I want to respond to all my clients' call within a reasonable time plus I can't ignore it if somebody does not have heating during the cold winter days. What if they get sick because they do not have heat?

Doctor: That's so admirable of you, Louis. Have you tried anything to relieve your pain?

Louis: Yes, I took Acetaminophen for my pain and I even applied an elastic bandage around my right elbow to support it, but it did not help to relieve the pain.

Doctor: Can you recall what makes the pain better or worse?

Louis: Obviously, if I work long hours with complicated plumbing problems, I feel much more severe pain on the end of the day.

Doctor: Do you have any swelling around your elbow?

Louis: Yes.

Doctor: Do you have numbness and tingling of hands or fingers? How about the strength of your handgrip?

Louis: I do not have any tingling or numbness of my hands and fingers, but my grip strength has gotten weaker. What is it?

Doctor: I think you might have developed "tennis elbow".

Louis: Tennis elbow? I do not play tennis. That's weird.

Doctor: We call it "tennis elbow" because it often occurs in people who play a lot of tennis. Actually, it is a repetitive use injury that results in inflammation of the extensor muscles of your elbow from repetitive use. You can have tennis elbow even if you do not play tennis sports. In your case, you have developed tennis elbow from repetitive use of your wrist and hand during wrist extension. Let me examine your arm. You will need rest, anti-inflammatory and analgesic medications and better support for your elbow while working.

49. Urinary Tract Infection

CC: "I have to go to the bathroom so often and I have a burning sensation when I pass my urine."

Redina is a 17 y/o Hispanic female who presented with urinary urgency, frequency and a burning sensation on urination for about 1 week. She could not sleep due to frequent trips to the bathroom last night, but when she tried to void, very little urine came out and after she void she has a sensation of incomplete emptying of her bladder. She also had a history of a UTI about 2 months ago. She is sexually active with her boyfriend, and they do not have any other sex partners. The patient used diaphragm for her birth control and Redina stated that sometimes she did not void after intercourse.

When she was diagnosed with a UTI 2 months ago, she did not visit her PCD for follow up care due to financial issues. She states that she does not have medical insurance and she thought she would be alright without seeing the doctor again since all her symptoms were resolved.

She has no history of STD in the past. She denied fever, chills, flank pain, headache, malaise or CVA tenderness (Costoverterbral angle tenderness). She denied vaginal discharge and no hematuria was reported. Before she developed these symptoms, she waited too long to urinate.

Because of the traffic, her bus did not arrive at her destination on its usual arrival time. She became restlessness in the bus, but she had no other choice. Her LMP was 2 weeks ago.

Dialogue

Doctor: Hi, Redina. You do not look good. How can I help you?

Redina: I have to go to the bathroom so often to void and when the feeling comes on, I have to run to the bathroom otherwise I'm afraid I might wet myself. It is very embarrassing.

Doctor: How long you have these symptoms?

Redina: For about 1 week.

Doctor: Any other symptoms? Have you had any UTI recently?

Redina: I get a burning sensation when I urinate and I had a UTI about 2 months ago.

Doctor: Did you go to your doctor for follow-up care?

Redina: …… (Pause) No. (Very quietly)

Doctor: Any reason you did not go?

Redina: Doctor, I do not have any health insurance. I just work part time and all my symptoms were went away after I finished the antibiotics. I thought I was cured. But I regret now that I did not go back to see my doctor again.

Doctor: OK, Are you sexually active?

Redina: Yes.

Doctor: How many partners do you have?

Redina: I have only one.

Doctor: How about your partner?

Redina: As far as I know, he does not have any other partners but me.

Doctor: After intercourse, do you usually void?

Redina: Sometimes yes, sometimes no. Last couple of times, I did not bother to go to the bathroom. It was just bothersome.

Doctor: Redina, do you use any type of birth control? How about your boyfriend?

Redina: My boyfriend refuses to use condom so I use a diaphragm for contraception.

Doctor: Did you have history of STD in the past?

Redina: No.

Doctor: OK. Do you have fever, chills, flank pain, headache or malaise?

Redina: No.

Doctor: How about a vaginal discharge or blood in the urine?

Redina: I do not have any discharge and my urine is OK .It's yellow.

Doctor: Redina, Can you recall whether you held your urine in a long time before you voided recently?

Redina: How did you know doctor? Actually, it happened once just before I noticed my symptoms. I was in the bus going home, but the bus arrived at my destination much later than I expected due to heavy traffic. I was so restless because I wanted to go to the bathroom, but there was nothing I could do but hold it in and wait.

Doctor: Redina, based on your history and symptoms, I think your UTI has recurred. Let me check your urine and do some blood test. I will prescribe antibiotics too. You will likely require a longer course of antibiotics this second time around as this infection may have developed resistance to the antibiotics that you were treated with previously. UTI's are common in women because of their short urethra and then as a result of sexual intercourse. In men, this is very uncommon and would prompt an evaluation for a structural problem. There, we would get an ultrasound of the kidneys, ureter and bladder looking for a blockage (obstruction) or stone in the kidney and bladder. In men, it may also require an evaluation by an urologist to check the prostate and possibly do a cystoscopy to examine the lining of the bladder wall for polyps, diverticula or cancer. The urologist may also want a transrectal ultrasound to more fully evaluate the prostate in men for cancer.

Reference List

Anthony M. Valeri MD.

Professor of Clinical Medicine Columbia University college of Physicians and Surgeons

Guide to Physical Examination and History Taking by Lynn S. Bickley 10th Edition

Primary Care Medicine by Goroll & Mulley 6th Edition

Current Diagnosis & Treatment Pediatrics by Hay, Levin, Sondheimer, Deterding

Pediatric Nurse Practitioner Certification Review by Fax & Gilman

Clinical Guidelines in Family Medicine 4th Edition by Uphold & Grabam

American Academy of Ophthalmology